Scheduling in Wireless Networks

Scheduling in Wireless Networks

Anna Pantelidou

Renesas Mobile Corporation
Elektroniikkatie 10
90570 Oulu
Finland
anna.pantelidou@renesasmobile.com

Anthony Ephremides

University of Maryland
College Park, MD 20742
USA
etony@umd.edu

now
the essence of knowledge

Boston – Delft

Foundations and Trends® in Networking

Published, sold and distributed by:
now Publishers Inc.
PO Box 1024
Hanover, MA 02339
USA
Tel. +1-781-985-4510
www.nowpublishers.com
sales@nowpublishers.com

Outside North America:
now Publishers Inc.
PO Box 179
2600 AD Delft
The Netherlands
Tel. +31-6-51115274

The preferred citation for this publication is A. Pantelidou and A. Ephremides, Scheduling in Wireless Networks, Foundation and Trends® in Networking, vol 4, no 4, pp 421–511, 2009

ISBN: 978-1-60198-420-3

Foundations and Trends® in Networking
Volume 4 Issue 4, 2009

Editorial Scope

Foundations and Trends® in Networking will publish survey and tutorial articles in the following topics:

- Ad Hoc Wireless Networks
- Sensor Networks
- Optical Networks
- Local Area Networks
- Satellite and Hybrid Networks
- Cellular Networks
- Internet and Web Services
- Protocols and Cross-Layer Design
- Network Coding
- Energy-Efficiency Incentives/Pricing/Utility-based
- Games (co-operative or not)
- Security
- Scalability
- Topology
- Control/Graph-theoretic models
- Dynamics and Asymptotic Behavior of Networks

Information for Librarians

Foundations and Trends® in Networking, 2009, Volume 4, 4 issues. ISSN paper version 1554-057X. ISSN online version 1554-0588. Also available as a combined paper and online subscription.

Foundations and Trends® in
Networking
Vol. 4, No. 4 (2009) 421–511

DOI: 10.1561/1300000030

Scheduling in Wireless Networks

Anna Pantelidou[1] and Anthony Ephremides[2]

[1] *Renesas Mobile Corporation, Elektroniikkatie 10, 90570 Oulu, Finland, anna.pantelidou@renesasmobile.com*
[2] *Institute for Systems Research, Department of Electrical and Computer Engineering, University of Maryland, College Park, MD 20742, USA, etony@umd.edu*

Abstract

We present a review of the problem of scheduled channel access in wireless networks with emphasis on ad hoc and sensor networks as opposed to WiFi, cellular, and infrastructure-based networks. After a brief introduction and problem definition, we examine in detail specific instances of the scheduling problem. These instances differ from each other in a number of ways, including the detailed network model and the objective function or performance criteria. They all share the "layerless" viewpoint that connects the access problem with the physical layer and, occasionally, with the routing layer. This review is intended to provide a reference point for the rich set of problems that arise in the allocation of resources in modern and future networks.

Contents

1

Introduction

This volume examines in some depth the fundamentals of the problem of scheduling transmissions over a multi-user shared channel. The origins of this problem are found in the area of Multiple Access (MA) [4, 35], where the traditional concept of simple "orthogonal" time-division [20] was enriched through the ideas of random access (known, more colloquially, as ALOHA [1, 32, 55]).

The original question was how to ensure the most productive use of the channel (that is, maximize the total, or "sum" throughput) when users have sporadic, "bursty", need to transmit and cannot coordinate their needs and actions amongst them. At the same time, the issue of sharing a channel was examined at a deeply theoretical level through an information-theoretic approach that aimed at determining the best "joint" rates at which different users can transmit over the shared channel if they can design their codebooks jointly and transmit without further coordination at the "protocol" level, that is without worrying about when to transmit. The reason for the latter, and very important, difference was that in the information-theoretic approach the sources were not assumed to have "bursty" and sporadic need for the channel but were backlogged and simply needed to transmit all the time. Of course, in that case, the receiver was assumed to be equipped with a multi-user detector (a concept that was formalized much later through the work of [71]) that was able to decode successfully the simultaneous transmissions of all users.

The first approach, which has been more characteristic of the work by what we call the "networking" community, has led to a large variety of protocols and standards that try to achieve the goal of maximum use of the channel through elaborate variations of practical exploitation of possibilities, such as feedback, carrier sensing, storing unsuccessful packets, often even relaying, etc. A large volume of literature exists that documents the efforts in this direction, e.g., [24, 42, 45, 56, 58, 59, 61, 66, 67, 68] to sample a few. The second approach has seen a similar voluminous body of work that has been mostly preoccupied with determination of the Shannon capacity region in variations of the shared channel model, such as pure multi-access, pure broadcast, relay, and interference models (see e.g., [21, 22, 27, 41, 62, 72]).

In both cases, the two approaches have fallen short of their ultimate goal due to the fundamentally complex nature of the problem. In the first case the major difficulty arises from the "dimensionality" issue, that is, the combinatorial nature of the problem, as the number of users increases. In the second case, the difficulty lies in the tremendous increase in analytical intractability of the Shannon-theoretic approach as the channel model becomes more complicated and/or the number of users increases.

The scheduling problem becomes more interesting and relevant to practice when the two approaches are partially blended. In particular, the first approach requires that transmissions from more than a single user cannot co-exist successfully in the same time slot. However, with multi-user detectors and by adjusting the transmission powers and the bit rates it is possible for several users to be successful simultaneously. Thus the question is what subset of users should be activated in each time slot of a frame. It must, of course, be taken into account that when more users are squeezed into the same slot their individual transmission rates must be reduced in order for them to tolerate each-others interference (and/or their transmission powers must be accordingly adjusted). Since the objective is to maximize bit/sec rates rather than the, not so informative, packets/slot rates, a clear trade-off emerges. Is it preferable for more users to transmit simultaneously but at reduced individual rates or is it better to time-share access by smaller sets of users that transmit, however, at higher individual rates? The answer is not clear and it depends on the specific environment, performance criterion, channel quality and gains, detector structure, modulation scheme, error control coding, etc.

In this volume, we follow this intermediate approach, more in the line of what the community has been referring to as "cross-layer" or "layerless" approach ([36, 40, 57, 63, 69]). Cross-layer approaches try to exploit linkages between the OSI layers, while layerless approaches consider, instead, the determination of variables of different layers simultaneously. In this volume we adopt the basic networking view of scheduling packet transmissions but, at the same time take into consideration the bit-rates that correspond to multi-user reception capability through physical-layer models. The purpose of this volume is to present a few samples of prior literature on the problem of scheduling and then outline through detailed illustrations some specific results that we have recently developed in addressing the problem in an innovative way. It is by no means exhaustive. The main two overarching ideas in our approach have been (i) to include physical-layer criteria in the determination of the probability of successful transmission and (ii) to opt for reducing the search space, rather than developing heuristics, in the case of combinatorial optimization problems. That is, in the "erasure" channel model (e.g., [19, 60, 70]), which has been gaining increasing attention recently and in which the probability of packet success is provided through a parameter value, we express this parameter value in terms of power, transmission bit-rate, and other variables from traditional physical layer communication-theoretic analyses. Furthermore, in the case of protocol of access optimization problems that are purely of integer programming nature, we do not follow the alternative of inventing heuristic sub-optimal solutions but, rather, we insist on rigorous optimization within the confines of a set of reduced solution space. We believe that the latter approach reveals insights that intelligent heuristics often fail to provide.

The first formulation of the problem of scheduling for efficient access to a shared channel, that we are aware of, appeared in [18]. A simple collision channel model had been considered but with the possibility of spatial re-use. That is, an "interference map" was assumed in terms of a graph that described all the independent sets of nodes in the graph, namely those sets of nodes that do not include "adjacent" nodes. The objective was to determine the shortest length of a frame of slots that would allow all nodes to transmit once in the frame without violating the "interference" constraint imposed by the interference rules on the graph. It was shown that the problem is NP-complete, and a distributed heuristic was developed that showed decent performance compared to the optimum that was computable in "small" instances of the problem.

This problem was revisited in more generality through a continuous approximation of the structure of the frame schedule in [26] where each interference-free set of nodes could be activated for an arbitrary amount of time and, again, the objective was the determination of the shortest duration of a schedule that would accommodate a given demand. It was shown that the continuous version of the problem could be solved in polynomial time, but there was no characterization of the optimal solution. A variation of this problem formulation that incorporated some physical-layer attributes was studied in [10, 11, 12].

Subsequently there has been a great deal of variations of these formulations that have focused mostly on heuristics and approximation ratios. A totally new attack to the problem of scheduling was developed in [66] where, again for a graph-based model of constraint-node-sets (i.e., sets of nodes that can transmit successfully at the same time), the objective was to determine not a schedule anymore, but, rather, a scheduling rule for these constraint-sets that guaranteed that if the input load to the network could be accommodated without excessive delays, then that rule of activation would assure that the delay objective would be met. The solution to this problem, which is intimately related to the notion of stable throughput region in a network, and that was the subject of the inaugural issue of the Foundations and Trends in Networking series [23], came to be known as the *back-pressure* algorithm and it has received extensive attention over the years with generalizations that include physical layer effects. In fact, the last section in this volume includes a particular generalization of the back-pressure result that introduces the notion of uncertainty in the knowledge of the channel state [52].

In what follows we formulate a precise model for the sharing of a common channel and examine several different variations of the scheduling problem that include different optimization criteria (e.g., proportional fairness), a minimum-time draining of the network with an initial load, and an asymptotically optimal policy determination for a multicast version of the problem. Due to the combinatorial nature of the scheduling problem, we also discuss a method for reducing the scheduling complexity. Finally, the general version of the back-pressure algorithm with imperfect channel state information (as alluded above) is presented and discussed.

The detailed work in the following sections represents research performed by the authors that led to the Ph.D. dissertation of Dr. Pantelidou and several recent journal publications and conference presentations. It is presented here in the general context of channel access and in an integrated and unified way.

1.1 Network Model

We consider a wireless network of M, possibly mobile, nodes each of which is equipped with a single transceiver (transmitter and receiver). We denote by the set $\mathcal{M} = \{1,\dots,M\}$ the set of all nodes in the network. We also denote by $\mathcal{L} = \{1,\dots,L\}$ the set of all links that can be potentially established among the M network nodes. The number of such links, L, can be as large as $M \times (M-1)$. We consider a slotted-time model where without loss of generality each slot t takes integral values, i.e., $t \in \{0,1,\dots\}$. At time slot t, each network node $n \in \mathcal{M}$ transmits at a power level $P_n(t)$. We denote by $\mathbf{P}(t)$ the vector of transmission powers at every network node, i.e., $\mathbf{P}(t) = (P_n(t), \forall n \in \mathcal{M})$. We assume that the power of the thermal noise is common for every node in the network. This assumption is non-restrictive and is made merely to simplify notation. We denote the power of the thermal noise by the variable N_0. Our results are valid also when the thermal noise is different at the various network nodes. It is often taken to be the same at all nodes for simplicity since it is not important and this will be the assumption we also make throughout this volume. The process $\{\mathbf{G}(t)\}_{t=0}^{\infty}$ defines the channel conditions between every pair of nodes in the network and it is assumed to change only at the beginning of a time slot t. Specifically, at time slot t the channel state $\mathbf{G}(t) = \{G_{(n,m)}(t),\ n,m \in \mathcal{M}\}$ gives the channel conditions between every pair of nodes $n,m \in \mathcal{M}$. We assume that the channel follows a block fading model with block length equal to the duration of a time slot. Hence, we assume that the channel conditions are allowed to change *only* at the beginning of each time slot and remain constant throughout the slot duration. The channel effects in our model can be due to node mobility, fading, pure path loss, etc. We assume that the channel process takes values in a set $\mathcal{G}$.

In certain sections of this volume we will assume that the channel is time-invariant. This is not only to make the solution of the scheduling instances tractable, but also to illustrate how the scheduling decisions are affected by the underlying channel conditions. Furthermore, in the rest of the volume, except for Section 5, we will ignore routing and assume single-hop networks where the M network nodes are separated in a set of sources of traffic, $\mathcal{T}$, and a set of destinations of traffic, $\mathcal{D}$, such that $\mathcal{M} = \mathcal{T} \cup \mathcal{D}$ and $\mathcal{T} \cap \mathcal{D} = \emptyset$. The single-hop network assumption, albeit simplifying, is interesting and highly non-trivial since it captures the fundamental problems that arise due to interference, when multiple nodes attempt simultaneous channel access.

Depending on the optimization criterion, in the rest of this volume we will assume three different cases, namely (a) sources that are *saturated* and always have data to transmit whenever they are activated, (b) sources with a *finite* amount of data traffic, and (c) sources with *bursty* arrivals. We will also consider three different traffic types, namely *unicast* traffic that originates from a single source and is destined to a single destination, *multicast* traffic that originates from a single source and is destined to multiple destinations, and *anycast* traffic that originates from a single source and is destined to any node within the set of destinations.

1.2 A Criterion for Successful Transmission

The fact that the wireless medium is shared by the network nodes poses limitations on the set of nodes that can concurrently transmit successfully.

In this volume, we incorporate these constraints on medium access through the Signal to Interference plus Noise Ratio (SINR) criterion. This model is of course approximate since it models the interference as Gaussian noise. However, it is intuitive, reasonable, and increasingly accurate as the number of interferers increases. We will say that a link exists when the transmission powers of the network nodes are given by the power vector $\mathbf{P}(t) = (P_n(t), n \in \mathcal{M})$ and when a node n transmits to node m (or that the transmission from n to m is successful) if the ratio of the received signal power to the sum of the thermal noise and the total interference at m exceeds a certain threshold γ_m, i.e.,

$$\mathrm{SINR}^{\mathbf{P}(t)}_{(n,m)}(t) := \frac{P_n(t)G_{(n,m)}(t)}{N_0 + \sum_{n'=1,n'\neq n}^{M} P_{n'}(t)G_{(n',m)}(t)} \geq \gamma_m. \qquad (1.1)$$

The exact value of the SINR threshold γ_m depends on various factors, such as the transmission rate, the target probability of bit error, the coding and modulation techniques employed at the transmission, etc. In this volume, we will only consider the dependence of this threshold on the transmission rate and assume that the rest of the parameters affecting it are fixed. It is well-known that the maximum transmission rate is an increasing function of the SINR threshold (see e.g., [25]). This gives rise to the following *trade-off*: By lowering the transmission rate, the corresponding value of the threshold decreases and hence more transmissions can jointly satisfy the condition of (1.1). On the other hand, by increasing the transmission rate the SINR threshold increases, thereby restricting the number of nodes that can concurrently access the channel successfully. Thus, it is not clear whether allowing more concurrent transmissions (less time-sharing) at lower rates is preferable to allowing fewer concurrent transmissions (more time-sharing) at higher rates. Shedding light in this trade-off will be one of the main objectives of this volume.

In this volume, we will consider this trade-off in a network where T sources of traffic wish to access the wireless medium. Under this setting, one extreme is to increase the threshold values so that only a single source can successfully transmit at any given time, that is, as in a Time Division Multiple Access (TDMA) fashion. Another extreme is to decrease the thresholds to the maximum values that allow all sources to successfully access the channel concurrently. In the latter case although more sources access the channel simultaneously, their rates will be significantly lower than the corresponding rates under TDMA operation. However, since they transmit continuously rather than in a TDMA fashion it is not clear how the long-term average rates of these two schemes compare to each other. These two extreme cases of operation will also be in the focus of this volume.

1.3 Organization of the Volume

The rest of the volume is organized as follows. In Section 2 we consider the minimum-length scheduling problem in single-hop wireless networks under unicast traffic. We present a rate control and scheduling policy that operates under the objective to empty a finite amount of traffic in the network queues in minimum time. We consider both the cases of static and time-varying networks. Next, in Section 3 we make a different assumption on the network traffic and performance criterion. In particular, we assume that the sources are sources of multicast traffic. Furthermore, we assume that they are saturated, that is, they always have data to send whenever they are activated. We obtain an on-line, gradient-based rate and power control algorithm that maximizes the overall network utility under this network setting. Since the scheduling problem is combinatorially complex, in the sequel, in Section 4 we present an approach to reduce complexity by restricting the set of possible scheduling and rate control decisions that the network control policy can take. In Section 5 we generalize the network topologies we consider and focus on general, multi-hop wireless networks under bursty arrivals. We consider the problem where the network control policies do not have perfect knowledge of the underlying channel conditions but take decisions only based on a, perhaps highly inaccurate, estimate. We introduce a class of policies that maximizes the stable throughput region of the network under channel uncertainty. Finally, in Section 6 we present our conclusions.

2

Minimum-Length Scheduling

In this section, we consider the problem of *minimum-length scheduling* in single-hop wireless networks. Specifically, we want to obtain a sequence of activations of wireless nodes so that a *finite* amount of unicast data traffic, residing at a subset of the nodes in the network reaches its intended destinations in *minimum time*. The problem of minimum length scheduling is closely related to the problems of network throughput or stable throughput maximization, since minimizing the time to deliver a fixed amount of data to its respective destinations, can be seen as maximizing the effective rate at which data traverses the network. However, it also differs from them in that some network resources become available to heavily loaded nodes when the lightly loaded ones become relieved. Furthermore, obtaining schedules of minimum length is a useful alternative metric that characterizes the traffic-carrying capabilities of wireless networks with non-stationary and non-ergodic channel variations, where the commonly used performance criteria of stable throughput and network capacity are not well defined. Although in this section we only focus on networks with stationary and ergodic channel behavior, we expect that a similar analysis can yield valuable insights regarding the more general case of non-ergodic and non-stationary wireless channels.

The topic of minimum-length scheduling has attracted a lot of attention recently [10, 11, 12, 26, 33, 34, 47]. In [26], the authors obtain a centralized, polynomial-time algorithm for static networks that finds a schedule of minimum-length satisfying a set of link traffic requirements. However, in [26] modeling of the physical layer is overly simplified as it is assumed that any two links can be successfully activated simultaneously as long as they do not share any common vertices. This simplification relates the minimum-length scheduling problem to the problem of obtaining a maximal matching in a non-bipartite graph [17]. However, due to the broadcast nature of the wireless medium *all* concurrent transmissions can potentially contribute to the total amount of interference at each receiver and, thus, cause a reception to fail.

In [12], the authors consider the problem of obtaining a schedule of minimum-length under the SINR interference model. They assume that the transmission rates are fixed and each transmitting node selects optimally its transmission power. In [12], the minimum-length scheduling problem is formulated as a linear program [9] that can possibly have a prohibitively large number of variables and thus is hard to solve. In [33] and [34] the authors consider the minimum-length scheduling problem for different sets of optimization parameters. Specifically, they consider the cases where (i) both the transmission powers and rates are fixed, (ii) the transmission powers can be optimized but the transmission rates are fixed, and (iii) the transmission powers are fixed and each transmitter is allowed to choose its rate from a predetermined, finite set of rates, common to all transmitters. In [33] and [34] the minimum-length scheduling problem is also formulated as a complex linear program, with a relatively small number of constraints and a large number of variables. To address the high complexity, the authors employ the technique of column generation [9], whose running time is faster, on average, than that of the original linear program [28].

Most of the prior work on the minimum-length scheduling problem assumes that the transmission rates are fixed. Due to the coupling between the physical layer and the medium access control in wireless systems, it is clear that a joint optimization of link activation and rate control will yield a better performance. We consider first static networks where the channel effect is due to pure path-loss. Also, we first assume a slotted-time model, and formulate the minimum-length scheduling problem as a shortest path between a given source–destination pair on a Directed Acyclic Graph (DAG). We obtain an optimal joint scheduling and rate control solution that provides a shortest path. Although finding a shortest path on a DAG has a polynomial complexity in the number of vertices and edges, this number grows exponentially as the size of the network and initial data traffic load increase. For this reason, we make the following simplifications. We first map the discrete-time problem to a continuous-time equivalent, where slots are eliminated. We then restrict the possible scheduling and rate control decisions to either communication "one at a time", in a TDMA fashion, or "all together". We explicitly characterize the optimal solution of this reduced problem. Understanding the behavior of the optimal policy, even for the reduced problem, is significant since it provides valuable intuition about when scheduling and rate control actions for one or the other extreme are preferable. This intuition, for example, can improve the performance of the column generation technique in [33] and [34] by providing the algorithm with those scheduling and rate control actions that are expected to be employed by an optimal policy in the reduced problem and thus improve that heuristic.

Furthermore, the cited work (i.e., [10, 11, 12, 26, 33, 34, 47]) studies the minimum-length scheduling problem *only* for time-invariant wireless networks. Since the wireless channel in reality is time-varying, in this section we will also address the case of time-varying channels. Our goal then becomes to find an optimal policy that minimizes the *expected time* required to deliver all the traffic to its respective destinations. We will solve the minimum-length scheduling problem by formulating it as a *Stochastic Shortest Path*, which is a special case of a Markov Decision Process (MDP). We will obtain an optimal scheduling and rate control policy through stochastic control methods. For time-invariant channel processes, this model reduces to finding a shortest path on a DAG and methods described in the first part of this section are applicable to compute the optimal solution.

2.1 Model Formulation

In this section, we restrict attention to single-hop, wireless networks where each source in the set of sources, $\mathcal{T}$, wishes to communicate with one distinct destination from the set of destinations, $\mathcal{D}$. To simplify notation we assume that the nth source wishes to communicate with the nth destination and that the cardinalities of the sets $\mathcal{T}$ and $\mathcal{D}$ are equal, that is, they satisfy $|\mathcal{T}| = |\mathcal{D}|$. We assume that $T = |\mathcal{T}|$. Simply speaking, this means that the network consists of T source and destination pairs as illustrated in Figure 2.1. The set of all source and destination pairs is also denoted by $\mathcal{T} = \{1,\dots,T\}$.

We also make the simplifying assumption that each source, n, can either transmit at its maximum transmission power $P_n^{\max}$ or remain silent. We denote the transmission power of the nth source at time slot t by $P_n(t)$, where $P_n(t) \in \{0, P_n^{\max}\}$.

We assume that each source of traffic has a *finite, fixed* amount of data units, e.g., a file of packets or bits to deliver to its corresponding destination. The objective is to activate the sources so that the time to deliver all the traffic to the intended destinations is minimized, i.e., to obtain a schedule of minimum length. Thus, source n has b_n bits to deliver to its corresponding destination. We denote by

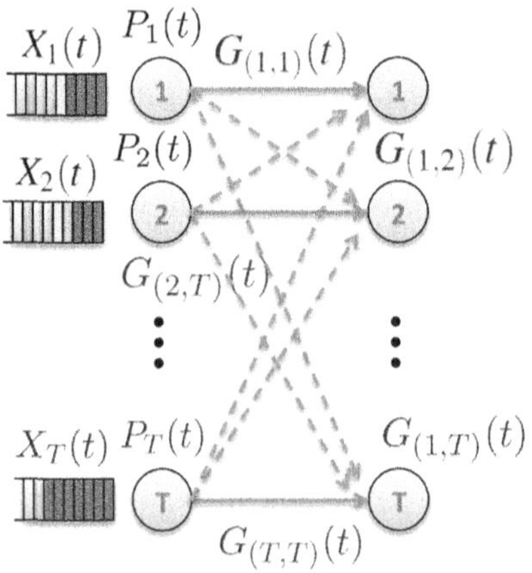

Fig. 2.1 A network of T source/destination pairs.

$\mathbf{b} = (b_1, \ldots, b_T)$ the vector of initial data traffic at each source. We also denote by $X_n(t)$ the queue size at source n at time slot t and by $\mathbf{X}(t) = (X_1(t), \ldots, X_T(t))$ the corresponding vector of queue sizes at all sources in the network. The queue size of each source at time slot 0 is equal to its initial data traffic, i.e., $\mathbf{X}(0) = \mathbf{b}$. The state space of the process $\{\mathbf{X}(t)\}_{t=0}^{\infty}$ is denoted by $\mathcal{X}$. Here, we also assume that the set of channel states $\mathcal{G}$ from which the channel process $\{\mathbf{G}(t)\}_{t=0}^{\infty}$ takes values is finite.

We consider joint scheduling and rate control policies, i.e., policies that at any given time slot jointly decide (a) which sources to activate and (b) their respective transmission rates. This information can be captured by the T-dimensional rate vector $\mathbf{r}(t) = (r_1(t), \ldots, r_T(t))$, where $r_n(t)$ is the rate of source n at slot t. Clearly, when a source n is not activated at time slot t, its corresponding rate $r_n(t)$ is equal to zero. In other words, a transmission rate vector implicitly specifies the scheduling decisions. The set of all feasible rate vectors contains those that are obtained by the following two-step procedure. We first identify all possible ways with which the network sources can be activated (by assigning to each source n either power 0 or $P_n^{\max}$) and then we assign to each subset of activated sources the maximum rates that allow them to jointly satisfy the SINR criterion of (1.1). Hence, in this section we allow the SINR thresholds to depend on the transmission rate. Thus, there exist $2^T - 1$ such T-dimensional transmission power vectors, each of which corresponds to an achievable rate vector. Naturally, for any given link, its highest instantaneous rate is achieved when it is scheduled alone due to the absence of interference. Furthermore, its rate is non-increasing as more and more links are concurrently activated. Clearly, the set of achievable rates depends on the current channel state $\mathbf{g} \in \mathcal{G}$. Hence, for every channel state $\mathbf{g}$, we denote by $\mathcal{R}(\mathbf{g})$ the finite, discrete set of feasible T-dimensional rate vectors. Then, the cardinality of $\mathcal{R}(\mathbf{g})$, i.e., $|\mathcal{R}(\mathbf{g})|$, is equal to $2^T - 1$ for every channel state $\mathbf{g} \in \mathcal{G}$.

In this section we are interested in obtaining optimal policies that take joint scheduling and rate control decisions under the objective of minimizing the (expected) time to deliver all data to the intended destinations. The policies we consider are aware of the network queue-sizes at all times. Furthermore, they are assumed to know the current channel conditions in order to make accurate scheduling decisions. For every slot t the pair of the channel state $\mathbf{G}(t)$ and queue sizes $\mathbf{X}(t)$ constitute the system state $\mathbf{S}(t)$. We denote by $\mathcal{S}$ the state space of the system state process $\{\mathbf{S}(t)\}_{t=0}^{\infty}$ which is given by

$$\mathcal{S} = \{(\mathbf{x},\mathbf{g})\colon \mathbf{x} \in \mathcal{X},\ \mathbf{g} \in \mathcal{G}\}. \tag{2.1}$$

We restrict our attention to *stationary* policies that make decisions merely based on the *current system state information.* Consider a state $i = (\mathbf{x},\mathbf{g}) \in \mathcal{S}$ and let the system state at time slot t satisfy $\mathbf{S}(t) = i$. Let us also define by the set $\mathcal{A}(i)$ the subset of the overall feasible scheduling and rate control decisions corresponding to state $i = (\mathbf{x},\mathbf{g}) \in \mathcal{S}$, i.e., $\mathcal{A}(i) \subseteq \mathcal{R}(\mathbf{g})$. Then, the policies we consider are given by the mapping

$$\mathbf{r}(t) = \boldsymbol{\pi}(i), \quad \boldsymbol{\pi}\colon \mathcal{X} \times \mathcal{G} \to \mathcal{A}(i) \subseteq \mathcal{R}(\mathbf{g}). \tag{2.2}$$

If it is a strict subset, scheduling will be suboptimal in general at the benefit of decreased complexity. Thus, in our model, we allow for the possibility of restricting the space of actions by limiting the range of allowable policies. Furthermore, by selectively choosing the elements of the set $\mathcal{A}(i)$ it is possible to obtain performance close to optimal while achieving considerable reduction in computational complexity.

We assume that every admissible policy uses the channel state information rationally so that a scheduled transmission must always be successful. Naturally, as reflected by the cardinality of the set $\mathcal{R}(\mathbf{g})$, the policies we consider are *non-idling*, i.e., they always activate at least one source that has a non-empty queue until all the queues in the network are empty. We call the class of stationary, non-idling policies given by the mapping (2.2) as *admissible* and denote them by $\mathbf{\Pi}$.

Consider a scheduling and rate control policy $\boldsymbol{\pi}$ that at every slot selects the transmission rates of all the sources. Then, the queue size process evolves according to the following equation

$$\mathbf{X}(t+1) = [\mathbf{X}(t) - \mathbf{r}(t)]^{+}, \quad t \geq 0 \tag{2.3}$$

where $[z]^{+} = \max\{z,0\}$.

Clearly, the queue size at each source n takes its maximum value at time slot 0, when it is equal to the initial demand b_n and, due to the absence of external arrivals, it keeps decreasing over time until it reaches zero. Under the above model, we proceed to formulate the minimum length scheduling problem for time-varying, and subsequently, static networks.

2.2 Time-Varying Networks

Throughout this section, we make the following assumption on the wireless channel process $\{\mathbf{G}(t)\}_{t=0}^{\infty}$.

Assumption 1. The channel process $\{\mathbf{G}(t)\}_{t=0}^{\infty}$ varies according to a stationary Markov Chain with transition probability from some channel state $\mathbf{g} \in \mathcal{G}$ to another channel state $\mathbf{g}' \in \mathcal{G}$ given by

$$p_{\mathbf{G}}(\mathbf{g}, \mathbf{g}') = \text{Prob}[\mathbf{G}(t+1) = \mathbf{g}' \mid \mathbf{G}(t) = \mathbf{g}], \quad \forall \mathbf{g}, \mathbf{g}' \in \mathcal{G}. \tag{2.4}$$

Due to the time variability of the channel process the length of the schedule, that we denote by N, is a random variable and thus "minimum-length" is meant "in the expected sense". This can be formulated as follows:

$$\textbf{minimize}: \quad \mathbb{E}[N] \tag{2.5}$$

$$\textbf{subject to}: \quad \mathbf{X}(N) = \mathbf{0}, \quad \mathbf{X}(0) = \mathbf{b}, \tag{2.6}$$

$$N \in \mathbb{N}, \tag{2.7}$$

where the expectation is taken with respect to the stationary probability distribution of the channel process.

We proceed to present a solution to the problem of (2.5)–(2.7) through stochastic control methods by considering admissible policies in the class $\mathbf{\Pi}$.

2.2.1 Stochastic Shortest Path Formulation

Since the wireless channel process $\{\mathbf{G}(t)\}_{t=0}^{\infty}$ is Markov and since the process of the queue sizes evolves according to (2.3), for every admissible policy, it is easy to show that the system process $\{\mathbf{S}(t)\}_{t=0}^{\infty}$ is also a Markov Chain, with state space $\mathcal{S}$ given by (2.1). We further define a subset $\mathcal{S}_{\text{term}}$ of the state space $\mathcal{S}$ to be the set of *terminating states* that correspond to empty queues, i.e.,

$$\mathcal{S}_{\text{term}} = \{(\mathbf{x},\mathbf{g})\colon \mathbf{x} = \mathbf{0},\ \mathbf{g} \in \mathcal{G}\}. \tag{2.8}$$

Evidently, from (2.3) this Markov Chain is absorbing. Therefore, from every non-terminating state a terminating state is reached with probability one in finite time under all admissible policies. Furthermore, once the system reaches any state in $\mathcal{S}_{\text{term}}$ it remains there forever. Hence, the objective becomes to reach a terminating state in minimum expected time by choosing the next state at every time slot. This will yield a schedule of minimum expected-length. This is a *Stochastic Shortest Path* problem, which is a special case of an MDP [7, 8].

The set of feasible scheduling and rate control actions corresponding to each system state $i = (\mathbf{x},\mathbf{g}) \in \mathcal{S}$ is the set $\mathcal{A}(i) \subseteq \mathcal{R}(\mathbf{g})$. The system is driven by the time-varying channel process $\{\mathbf{G}(t)\}_{t=0}^{\infty}$. Taking an action leads to different states with different probabilities depending on the evolution of the channel process unless the system has already reached a terminating state.

Let $p_{\mathbf{r}}(i,j)$ be the transition probability from system state $i = (\mathbf{x},\mathbf{g})$ to state $j = (\mathbf{x}',\mathbf{g}')$ by taking action $\mathbf{r} = \boldsymbol{\pi}(\mathbf{x},\mathbf{g}) \in \mathcal{A}(i)$. Then we have:

$$\begin{aligned} &p_{\mathbf{r}}(i,j) \\ &\quad = \text{Prob}[\mathbf{X}(t+1) = \mathbf{x}', \mathbf{G}(t+1) = \mathbf{g}' \,|\, \mathbf{X}(t) = \mathbf{x}, \mathbf{G}(t) = \mathbf{g}, \boldsymbol{\pi}(\mathbf{x},\mathbf{g}) = \mathbf{r}]. \end{aligned}$$

From (2.3) and Assumption 1 it is easy to see that $p_{\mathbf{r}}(i,j)$ can be written as

$$p_{\mathbf{r}}(i,j) = \begin{cases} p_{\mathbf{G}}(\mathbf{g},\mathbf{g}'), & \text{if } (\mathbf{x}-\mathbf{r})^{+} = \mathbf{x}',\ i,j \in \mathcal{S} \\ 0, & \text{otherwise.} \end{cases} \tag{2.9}$$

Note that from the Markovianess of the channel process and the admissibility of the policy $\boldsymbol{\pi}$, the transition probability $p_{\mathbf{r}}(i,j)$ is time invariant and does not depend on the previous system states.

We define the cost of taking action $\mathbf{r}$ and going from state i to state j as $\tilde{c}_{\mathbf{r}}(i,j)$. For every system state i, action $\mathbf{r} \in \mathcal{A}(i)$, and system state j such that $p_{\mathbf{r}}(i,j) > 0$ we assume that $\tilde{c}_{\mathbf{r}}(i,j) = 1$. This represents the fact that in order to go from state i to state j by taking this action one needs to spend one time slot. Let us further define the *cost per stage* $c_{\mathbf{r}}(i)$ to be the expected cost when control action $\mathbf{r} \in \mathcal{A}(i)$ is chosen while being at a non-terminating state $i \in \mathcal{S} \backslash \mathcal{S}_{\text{term}}$. It is clear that $c_{\mathbf{r}}(i) = \sum_{j \in \mathcal{S}} p_{\mathbf{r}}(i,j)\tilde{c}_{\mathbf{r}}(i,j) = 1$. Once a terminal state $i \in \mathcal{S}_{\text{term}}$ is reached, no more cost is incurred and the system remains there forever, i.e., $c_{\mathbf{r}}(i) = 0$, $\forall\, \mathbf{r} \in \mathcal{A}(i)$, $i \in \mathcal{S}_{\text{term}}$.

2.2.2 An Optimal Policy

Let $\mathcal{N}_{\boldsymbol{\pi}}(i)$ be the expected time to empty the queues in the network starting from state i under a policy $\boldsymbol{\pi} \in \boldsymbol{\Pi}$. Clearly, $\mathcal{N}_{\boldsymbol{\pi}}(i) = 0$ under any policy $\boldsymbol{\pi} \in \boldsymbol{\Pi}$ for every terminating state $i \in \mathcal{S}_{\text{term}}$. Then the minimum expected schedule length starting at state i, $\mathcal{N}^{\star}(i)$, is given by

$$\mathcal{N}^{\star}(i) = \min_{\boldsymbol{\pi} \in \boldsymbol{\Pi}} \mathcal{N}_{\boldsymbol{\pi}}(i), \quad \forall i \in \mathcal{S} \setminus \mathcal{S}_{\text{term}}.$$

A policy $\boldsymbol{\pi}^{\star}$ is optimal if it achieves the minimum $\mathcal{N}^{\star}(i)$ for every non-terminating state $i \in \mathcal{S} \setminus \mathcal{S}_{\text{term}}$, i.e.,

$$\mathcal{N}_{\boldsymbol{\pi}^{\star}}(i) = \mathcal{N}^{\star}(i), \quad \forall i \in \mathcal{S} \setminus \mathcal{S}_{\text{term}}.$$

To solve the above shortest path problem, two commonly used methods are policy iteration and value iteration [8]. Due to the large state space of the problem, value iteration is easier to compute and, hence, will be used here. Consider the value iteration algorithm and the corresponding "expected" time $\bar{\mathcal{N}}_k(i)$ to empty the queues starting from state i at the kth iteration. Assume that $\bar{\mathcal{N}}_0(i) = \infty$ for all states $i \in \mathcal{S}$. We borrow the following properties from [8].

Lemma 2.1. The value iteration method converges to the optimal cost function, i.e.,

$$\mathcal{N}^{\star}(i) = \lim_{k \to \infty} \bar{\mathcal{N}}_k(i), \quad \forall i \in \mathcal{S} \setminus \mathcal{S}_{\text{term}},$$

where

$$\bar{\mathcal{N}}_{k+1}(i) = 1 + \min_{\mathbf{r} \in \mathcal{A}(i)} \sum_{j \in \mathcal{S}} p_{\mathbf{r}}(i,j) \bar{\mathcal{N}}_k(j), \quad i \in \mathcal{S} \setminus \mathcal{S}_{\text{term}}.$$

Lemma 2.2. The optimal solution to a Stochastic Shortest Path problem must satisfy Bellman's equation, i.e., for every non-terminating state $i \in \mathcal{S} \backslash \mathcal{S}_{\text{term}}$ it must be true that

$$\mathcal{N}^{\star}(i) = 1 + \min_{\mathbf{r} \in \mathcal{A}(i)} \left\{ \sum_{j \in \mathcal{S}} p_{\mathbf{r}}(i,j) \mathcal{N}^{\star}(j) \right\}.$$

Hence, the optimal scheduling and rate control policy $\boldsymbol{\pi}^{\star}$ for every state $i \in \mathcal{S} \setminus \mathcal{S}_{\text{term}}$ is given by

$$\boldsymbol{\pi}^{\star}(i) = \arg\min_{\mathbf{r} \in \mathcal{A}(i)} \left\{ \sum_{j \in \mathcal{S}} p_{\mathbf{r}}(i,j) \mathcal{N}^{\star}(j) \right\}, \quad \forall i \in \mathcal{S} \setminus \mathcal{S}_{\text{term}}.$$

Although the value iteration method optimally solves the aforementioned stochastic shortest path problem, in general it may require an infinite number of iterations until it converges. However, if the Markov Chain of the system evolution is acyclic, then it was shown in [8] that the value iteration method for each state converges in a finite number of iterations, which are at most as many as the non-terminating states of the Markov Chain.

It is easy to see that the Markov Chain driving our system is acyclic. This is because starting from one of the states i whose queue size satisfies $\mathbf{X}(0) = \mathbf{b}$, the queue sizes in the network are non-increasing with time as given in (2.3) under any admissible policy. Thus, termination in finite time is guaranteed under our network model.

2.3 Static Networks

In this section, we restrict our attention to static networks, where the channel qualities $G_{(n,m)}(t)$ are time-invariant, i.e., we ignore effects of fading or user mobility. Thus, the cardinality of the set $\mathcal{G}$ is equal to one. To simplify notation, in this section we denote the channel quality $G_{(n,m)}(t)$ as $G_{(n,m)}$. Furthermore, since there is a single channel state $\mathbf{g}$, to simplify notation, we will write $\mathcal{R}$ to denote $\mathcal{R}(\mathbf{g})$ for $\mathbf{g} \in \mathcal{G}$ and $\mathcal{A}$ to denote $\mathcal{A}(i)$ for $i = (\mathbf{g}, \mathbf{x})$, $\mathbf{g} \in \mathcal{G}$. We consider a scheduling and rate control policy that at every time slot t identifies a rate vector $\mathbf{r}(t) = (r_1(t), \ldots, r_T(t)) \in \mathcal{A} \subseteq \mathcal{R}$. The rate vector $\mathbf{r}(t)$ specifies which sources are activated at time slot t and their respective rates.

Since there is no randomness in the channel state, the length of the schedule is non-random. Thus, the minimum-length scheduling problem can be written as follows:

$$\textbf{minimize}: \quad N \tag{2.10}$$

$$\textbf{subject to}: \quad \mathbf{X}(N) = \mathbf{0}, \quad \mathbf{X}(0) = \mathbf{b}, \tag{2.11}$$

$$N \in \mathbb{N}. \tag{2.12}$$

In the specific case of pure TDMA scheduling, combined with rate control, where only a single source can be active at any given time, the solution of the above problem becomes trivial. Specifically, each source must be active for as many time slots as needed to empty its queue. The required number of such time slots for each source n is equal to the ratio of its initial demand b_n divided by its corresponding rate when it accesses the channel individually. Then, the minimum total time that is needed until all the queues are empty is equal to the sum of the time slots required by each source. The order in which the sources must be activated is immaterial; they can be chosen in a round-robin or random fashion.

However, the solution of the optimization problem given by (2.10)–(2.12) is in general a non-trivial discrete optimization problem. In the following sections, we provide an optimal graph-theoretic algorithm by mapping it to a shortest path problem on a DAG.

2.3.1 The Equivalent DAG Representation

We construct a weighted DAG, $G = (V, E)$, as follows: we assume that every vertex $\mathbf{u} \in V$ of the DAG represents a queue-size vector that can be obtained, starting from a vector of queues $\mathbf{X}(t)$, by employing some scheduling and rate control action chosen from the set $\mathcal{A}$. Furthermore, every directed edge $(\mathbf{u}, \mathbf{v}) \in E$ represents one such action in $\mathcal{A}$. We say that the edge $(\mathbf{u}, \mathbf{v})$ is *incident from* $\mathbf{u}$ and *incident to* $\mathbf{v}$. Hence, from every vertex $\mathbf{x}^i$ we can have $|\mathcal{A}|$ edges that are incident from $\mathbf{x}^i$, each corresponding to a different rate vector $\mathbf{r}^i, i = 1, \ldots, |\mathcal{R}|$. Each such edge is incident to a graph node $\mathbf{y}^i = [\mathbf{x}^i - \mathbf{r}^i]^+$. We disallow those edges that correspond to rate vectors which activate sources with empty queues. Therefore, the actual number of edges that are incident from a vertex can be less than $|\mathcal{A}|$. The weight of each edge $(\mathbf{u},\mathbf{v})$ is equal to one to indicate that we need one time slot to go from vertex $\mathbf{u}$ to vertex $\mathbf{v}$. From now on, we will refer to action $\mathbf{r}^i$ by means of the edge $(\mathbf{x}^i,\mathbf{y}^i)$. The unique source node $\mathbf{x}^0$ of the DAG corresponds to the vector of initial demands, $\mathbf{X}(0)$.

In Figure 2.2 we give a toy example of such a graph for a network of two source and two destination pairs. Note that all the units can be scaled appropriately to represent meaningful queue sizes and transmission rates. We assume that the initial demands are $b_1 = 4$ bits and $b_2 = 6$ bits and that we have three possible scheduling and rate control actions: (i) only source 1 accesses the channel at a rate of 3 bits/s, (ii) only source 2 accesses the channel at a rate of 3 bits/s, and (iii) both sources concurrently transmit at a rate of 2 bits/s each. For simplicity we assume that the duration of a slot is equal to 1 s. Figure 2.2 depicts the DAG that is obtained by these three actions. Note that from each vertex all the three rate control actions are allowed, as long as each action schedules sources with non-empty queues. For example,

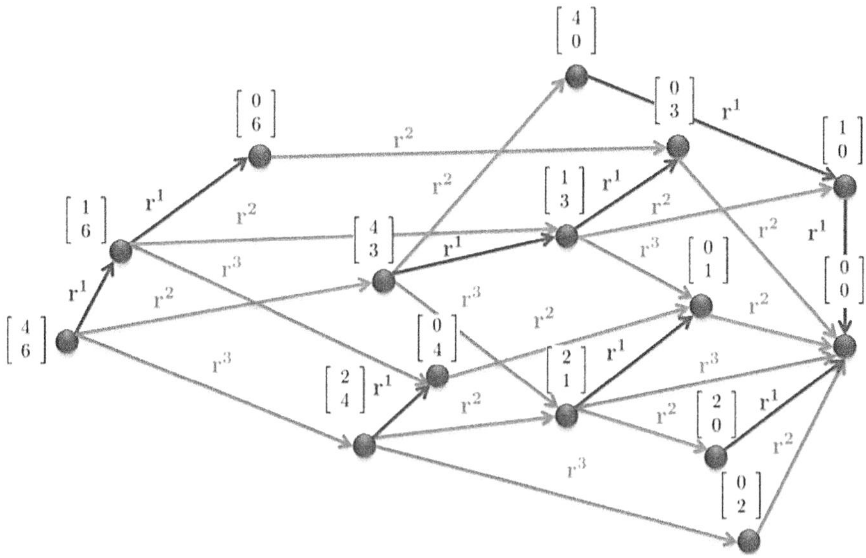

Fig. 2.2 A DAG construction corresponding to initial demands $b_1 = 4$ bits and $b_2 = 6$ bits and three rate control actions $\mathbf{r}^1 = [3,0]$, $\mathbf{r}^2 = [0,3]$, and $\mathbf{r}^3 = [2,2]$.

in Figure 2.2 the only viable rate control action for the queue-size vector [4,0] is to activate source 1 individually.

As we observe from Figure 2.2 for any path of vertices $\langle \mathbf{x}^0, \mathbf{x}^1, \mathbf{x}^2, \ldots, \mathbf{x}^{\bar{p}} \rangle$ the queue-size vector of each vertex in the path has to be component-wise larger or equal to the queue size of any other vertex that succeeds it in the path and the queue-size vectors of any two vertices on the graph cannot be the same. As a result, the overall graph is directed and acyclic. Finally, it is clear that every path starting at the source $\mathbf{s}$ ends at the $\mathbf{0}$-vector. Moreover, the weight of any sub-path $\langle \mathbf{x}^0, \mathbf{x}^1, \mathbf{x}^2, \ldots, \mathbf{x}^{\bar{p}} \rangle$ is equal to its length $\bar{p}$, which is effectively the number of time slots required to go from $\mathbf{x}^0$ to $\mathbf{x}^{\bar{p}}$ along the specified path, since by construction of this DAG the weight of each edge is equal to one time slot. Thus, the initial problem given by (2.10)–(2.12) is transformed into a shortest path problem on a weighted single-source DAG.

For the example given in Figure 2.2, the shortest path algorithm selects the sequence of actions $\mathbf{r}^3, \mathbf{r}^3, \mathbf{r}^2$ and the minimum schedule length is equal to 3 s (slots). Note that the sequences of actions $\mathbf{r}^2, \mathbf{r}^3, \mathbf{r}^3$ and $\mathbf{r}^3, \mathbf{r}^2, \mathbf{r}^3$ are also optimal as the order in which the actions are taken is immaterial in terms of minimizing the time needed to empty the queues, under the assumption of static channels. Also, it is worth mentioning that the length of the optimal schedule obtained through rate control is, naturally, no longer than that of TDMA (employing only actions $\mathbf{r}^1$ and $\mathbf{r}^2$) which, in this example, has length 4. Furthermore, it is reasonable to expect that the difference in the length of the schedule under the two schemes can become significant when the number of source/destination pairs in the network increases or when the values of initial demands are large. This will also be illustrated through a set of numerical results in Section 2.4.

Finding a shortest path on a DAG has running time equal to $\Theta(|V| + |E|)$ [17]. Hence, the number of operations needed to compute a shortest path on a single-source DAG is of polynomial complexity in the number of vertices and edges. However, in our DAG construction this number grows exponentially (i) in the number of sources when $\mathcal{A} = \mathcal{R}$ since from every vertex there exist $2^T - 1$ potential edges that are incident from it and (ii) as the initial demands increase. Therefore the overall complexity of the algorithm becomes exponential and is rendered impractical. Later, in Section 4 we will discuss a method to reduce the complexity of the optimal solution by compromising optimality.

2.4 Numerical Results

In this section, we illustrate our analytical results through a few numerical calculations. We consider a network of two source/destination pairs as in Figure 2.1 with $T = 2$. The channel process $\{\mathbf{G}(t)\}_{t=0}^{\infty}$ is Markov and switches between two states, namely a *good* state, $\mathbf{G}$, and a *bad* state, $\mathbf{B}$. When the channel is in the good state, both sources have channels of good quality to their destinations; otherwise both channels are bad. The transition probabilities of this Markov Chain are shown in Figure 2.3.

Since we have two source/destination pairs, there exist three possible rate vectors corresponding to each channel state, denoted by $\mathbf{r}^n(\mathbf{g})$, $n = 1,2$ when only the nth source is activated and by $\mathbf{r}^3(\mathbf{g})$ when both sources are activated, under channel state $\mathbf{g} \in \{\mathbf{B},\mathbf{G}\}$. We first assume that the initial demands are $b_1 = 4$ bits and $b_2 = 6$ bits which is the case discussed in Section 2.3. We consider three scenarios associated with different achievable rates for each channel state.

- **Scenario 1:** In this case we assume that the channel realization is immaterial. When the nth source $n = 1,2$ is activated alone, under either the good or the bad channel state, its achievable rate is 3 bits/s and when both sources are activated simultaneously, the corresponding rates are 2 bits/s for each. In this case, the minimum expected time to empty the queues is 3 s (slots), i.e., equal to the result of the static network example of Section 2.3.

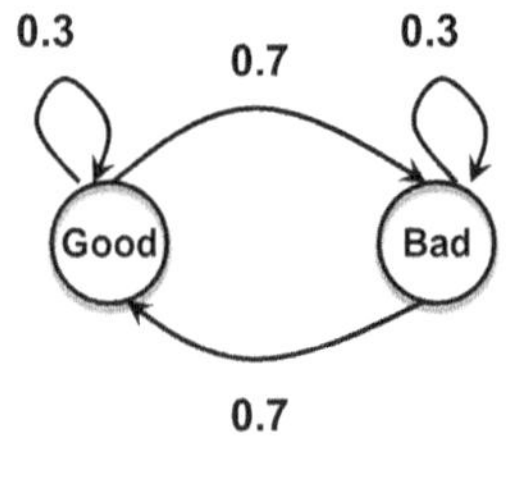

Fig. 2.3 A two-state Markovian channel representation.

- **Scenario 2:** Under the good channel state, **G**, the achievable rates are equal to those of the case of Scenario 1, i.e., when the nth source, $n = 1, 2$ is activated alone its achievable rate is 3 bits/s and when both sources are activated simultaneously, the corresponding rates are 2 bits/s for each source. However, under the bad channel, **B**, the achievable rates are strictly worse (2 bits/s for individual transmission and 1 bit/s for each source under concurrent transmission). Naturally, we anticipate that the expected time required to empty the queues is more than 3 s (slots).
- **Scenario 3:** We assume that under the bad channel state, **B**, the achievable rates are equal to the ones in Scenario 1, but the good channel is better and thus allows higher rates (4 bits/s when a source is activated individually and 3 bits/s when they are both activated simultaneously). Naturally, the expected time to empty the queues will decrease to a value less than 3 s (slots).

The above three scenarios and the corresponding minimum expected schedule lengths are shown in Table 2.1, where we have assumed that

Table 2.1. Expected time required to empty the queues for different values of initial demands, under Scenarios 1–3, assuming that the channel starts from a good state.

	Good Channel			Bad Channel			
Demand	$\mathbf{r}^1(\mathbf{G})$	$\mathbf{r}^2(\mathbf{G})$	$\mathbf{r}^3(\mathbf{G})$	$\mathbf{r}^1(\mathbf{B})$	$\mathbf{r}^2(\mathbf{B})$	$\mathbf{r}^3(\mathbf{B})$	$\mathbb{E}[N]$
	$\begin{bmatrix}3\\0\end{bmatrix}$	$\begin{bmatrix}0\\3\end{bmatrix}$	$\begin{bmatrix}2\\2\end{bmatrix}$	$\begin{bmatrix}3\\0\end{bmatrix}$	$\begin{bmatrix}0\\3\end{bmatrix}$	$\begin{bmatrix}2\\2\end{bmatrix}$	3.00
$\begin{bmatrix}4\\6\end{bmatrix}$	$\begin{bmatrix}3\\0\end{bmatrix}$	$\begin{bmatrix}0\\3\end{bmatrix}$	$\begin{bmatrix}2\\2\end{bmatrix}$	$\begin{bmatrix}2\\0\end{bmatrix}$	$\begin{bmatrix}0\\2\end{bmatrix}$	$\begin{bmatrix}1\\1\end{bmatrix}$	3.62
	$\begin{bmatrix}4\\0\end{bmatrix}$	$\begin{bmatrix}0\\4\end{bmatrix}$	$\begin{bmatrix}3\\3\end{bmatrix}$	$\begin{bmatrix}3\\0\end{bmatrix}$	$\begin{bmatrix}0\\3\end{bmatrix}$	$\begin{bmatrix}2\\2\end{bmatrix}$	2.79
	$\begin{bmatrix}3\\0\end{bmatrix}$	$\begin{bmatrix}0\\3\end{bmatrix}$	$\begin{bmatrix}2\\2\end{bmatrix}$	$\begin{bmatrix}3\\0\end{bmatrix}$	$\begin{bmatrix}0\\3\end{bmatrix}$	$\begin{bmatrix}2\\2\end{bmatrix}$	50.00
$\begin{bmatrix}100\\100\end{bmatrix}$	$\begin{bmatrix}3\\0\end{bmatrix}$	$\begin{bmatrix}0\\3\end{bmatrix}$	$\begin{bmatrix}2\\2\end{bmatrix}$	$\begin{bmatrix}2\\0\end{bmatrix}$	$\begin{bmatrix}0\\2\end{bmatrix}$	$\begin{bmatrix}1\\1\end{bmatrix}$	66.86
	$\begin{bmatrix}4\\0\end{bmatrix}$	$\begin{bmatrix}0\\4\end{bmatrix}$	$\begin{bmatrix}3\\3\end{bmatrix}$	$\begin{bmatrix}3\\0\end{bmatrix}$	$\begin{bmatrix}0\\3\end{bmatrix}$	$\begin{bmatrix}2\\2\end{bmatrix}$	40.25

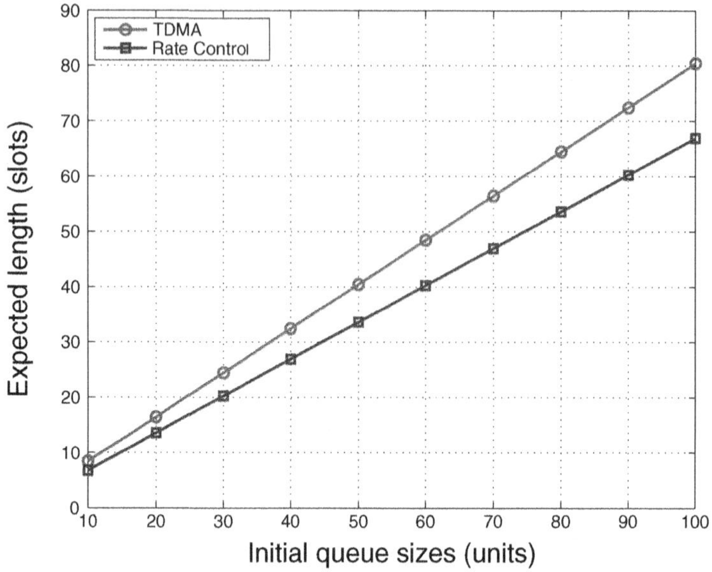

Fig. 2.4 Performance comparison of the expected schedule length achieved by the optimal policy and by TDMA scheduling.

the channel starts from a good channel state. Similar results were observed for higher initial demands ($b_1 = 100$ bits and $b_2 = 100$ bits), which are also illustrated in Table 2.1. Note that although time is measured in terms of time slots, the average minimum expected length takes fractional values since it is computed by taking the expectation with respect to the probability distribution of the channel state.

Next, in Figure 2.4 we illustrate the performance comparison between the optimal policy and a pure TDMA scheme that activates only a single source at a time. Specifically, we consider the same single-hop network of two source/destination pairs discussed above under Scenario 2. We plot the expected schedule-length for the above two schemes as a function of the values of the initial data traffic. For simplicity, the initial queue sizes at each source are assumed to be equal. As expected, we observe that the difference between the expected time to empty the queues under the optimal policy and under the TDMA scheme increases as the initial queue sizes increase. This result confirms that permitting concurrent transmissions through rate control can provide considerable gains.

3

Utility Maximization

In this section, we consider the problem of utility maximization for time-varying, single-hop wireless networks under multicast traffic. We consider a set of multicast sources that are always backlogged (that is we do not consider bursty arrivals, queuing, and questions of stability). The sets of destinations in different multicast sources may be overlapping. The objective is to schedule the sources by selecting their instantaneous transmission rates and powers so that the average values of the rates at the destinations maximize the sum of the destination utilities.

The problem of joint scheduling and rate control for maximizing the user utility has been studied extensively in the literature. A large body of work focuses on scheduling of the downlink channel of a base station transmitting *unicast* data traffic to a set of mobile terminals. One particular example is the Proportional Fair Sharing scheduler (PFS) introduced by Qualcomm. The PFS selects a single terminal at any given time, according to a TDMA-based scheme, the one that maximizes the ratio of a user's instantaneous rate to the average rate it has received so far in order to achieve proportional fairness. Therefore, those terminals that received comparably lower average data rates until the current decision instant are more likely to be selected in the optimal solution [2, 37, 38].

However, as it is shown in a variety of settings [6, 48, 49], TDMA scheduling of a set of nodes one at a time is not optimal. In fact, it is shown in [48] that in a static wireless network, depending on the channel conditions, it may be beneficial to allow all the nodes to operate concurrently for a certain period of time under the objective of proportional fairness for unicast traffic. Similar results are obtained in [49] for the case of multicast traffic.

The non-optimality of TDMA is especially true when the cost criteria, like that of proportional fairness, are more elaborate and intricately dependent on the transmission powers. A brief summary of relevant work includes [3], where the authors consider the problem of rate control for unicast traffic in time-varying wireless networks. Their formulation permits the scheduling of concurrent unicast transmissions. The authors introduce an optimal rate-control policy under the objective of maximizing a user utility that is a strictly concave, continuously differentiable, and increasing function of the average rate of each destination. Furthermore, the channel process considered in [3] is assumed to be stationary and ergodic. A subsequent study, [64], considers the problem of optimal rate allocation for a switch serving a set of queues under the objective of utility maximization. Although a broader class of utility functions is considered in [64], the switch is restricted to change states according to a finite-state, stationary and ergodic Markov Chain.

In all these cases the rate-control problem for utility maximization was limited to unicast traffic. However, network traffic is often multicast. The problem of multicasting to a multicast group over a wireless network has been studied under a variety of contexts [15, 16, 29, 44, 76]. In [74] the authors consider a base station with multicast traffic to a number of user groups. It is assumed that only a single multicast group can be chosen to receive at any given time and that all the terminals in the multicast group receive at a common rate (i.e., single-rate multicast). The multicast scheduler needs to decide which *unique* group to serve and at which *rate*, under two performance objectives. First, the authors in [74] consider the objective of proportional fairness with respect to the total rate that each multicast group receives. Next, they also consider the objective of total rate for each terminal when the latter is a member of multiple multicast groups. Recently, in [46], we considered the problem of utility maximization for *multicast* traffic in time-varying wireless networks (not limited to single base station architectures), through joint rate and power control by permitting simultaneous node activations.

3.1 Model Formulation

3.1.1 Network Model

In this section, we consider time-varying, single-hop wireless networks. Specifically, we consider networks that consist of T sources and D destinations and we denote by $\mathcal{T} = \{1, 2, \ldots, T\}$ and $\mathcal{D} = \{1, 2, \ldots, D\}$ the sets of sources and destinations in the network respectively. Although in Section 2 we assumed unicast traffic, in this section we will consider the more general case of multicast. Thus, each source $n \in \mathcal{T}$ is associated with a *multicast* session and multicasts traffic to a set of destinations $\mathcal{D}(n) \subseteq \mathcal{D}$. We denote by $|\mathcal{D}(n)|$ the cardinality of the set $\mathcal{D}(n)$. Our model captures the special cases of *unicast* ($|\mathcal{D}(n)| = 1$) and *broadcast* traffic ($|\mathcal{D}(n)| = D$). We assume that different multicast sessions may have overlapping destination sets, i.e., for any two sources $n, n' \in \mathcal{T}$ it is possible that $\mathcal{D}(n) \cap \mathcal{D}(n') \neq \emptyset$. As an example, in Figure 3.1 both sources 1 and 2 multicast to destination 1. Another difference from Section 2 is that here we assume that each multicast source is always backlogged, that is, it always has data to send whenever it is activated. This traffic model has to be distinguished from other alternatives that assume burstiness in traffic and that focus on stability.

We assume that the vector of transmission powers takes values from a continuous, compact set $\mathcal{P}$ of allowable power values, i.e., $\mathbf{P}(t) \in \mathcal{P}$. The compactness of the set of powers imposes a maximum instantaneous power constraint. We also make the assumption that the channel process $\{\mathbf{G}(t)\}_{t=0}^{\infty}$ is *stationary* and *ergodic* and it takes values from a continuous set $\mathcal{G}$ with a probability density function $f_{\mathcal{G}}(\cdot)$. The above are illustrated in Figure 3.1.

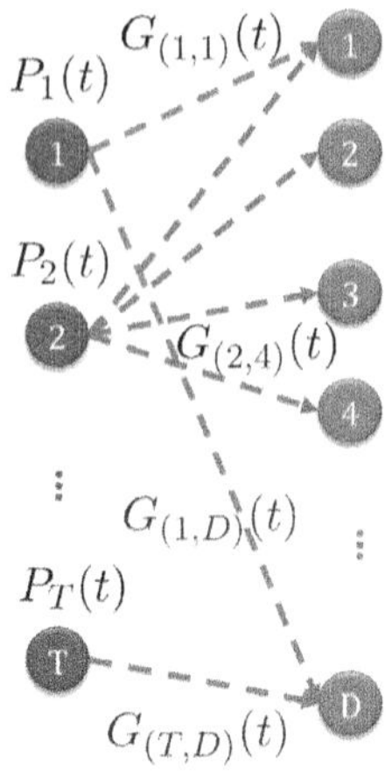

Fig. 3.1 A network of T multicast sources and D destinations.

We consider two types of destinations: (i) destinations that can only receive from a single source at any given time, i.e., Single Packet Reception (SPR), and (ii) destinations with Multi-Packet Reception (MPR) capabilities. Under MPR a destination may successfully receive concurrently from multiple sources as long as the SINR from each one of them exceeds the required threshold. Hence, under MPR two (or more) multicast sources with overlapping receiving nodes can concurrently transmit successfully, while in the SPR case only a single transmission can be received successfully at any given time.

Under unicast traffic, throughput is unambiguously defined as the rate at which data is delivered to a destination successfully. However, this is not as simple under multicast traffic. Shall we measure multicast throughput based on the messages sent by the source or based on the messages received by the destinations? Defining throughput in terms of the transmission rate of a multicast source would give two sources operating at the same rate equal weights, regardless of the number of destinations to which each of them transmits. In this volume, we will measure throughput with respect to the messages received by the destinations.

The transmission rate and power of each source $n \in \mathcal{T}$ at time slot t are denoted by $r_n(t)$ and $P_n(t)$ respectively. Let $\mathbf{r}(t) = (r_n(t), n \in \mathcal{T})$

and $\mathbf{P}(t) = (P_n(t), n \in \mathcal{T})$ be the corresponding transmission rate and power vectors. We will define a successful transmission according to the SINR criterion of (1.1), where the potential nodes causing interference are the sources in $\mathcal{T}$ (and thus we set $M = T$ in (1.1)). Specifically, $\forall n \in \mathcal{T}$ we will say that source n *multicasts* successfully at slot t at rate $r_n(t)$, if the SINR at each destination $d \in \mathcal{D}(n)$ exceeds the required threshold, i.e., if

$$\mathrm{SINR}_{(n,d)}^{\mathbf{P}(t)}(t) \geq \gamma_d(r_n(t)), \quad \forall d \in \mathcal{D}(n). \tag{3.1}$$

Here, we focus on the dependence of the threshold on the transmission rate. We will write $\gamma_d(r_n(t))$ to indicate the SINR threshold at time slot t that represents the minimum value of SINR that allows successful transmission when source n transmits at rate $r_n(t)$ to destination d. Note that here we consider the general case where different destinations may employ different modulation and coding schemes and may have different bit-error-rate requirements. This is captured in our model by permitting the SINR threshold functions $\gamma_d(r_n(t))$ to be dependent on the identity of the particular destination d.

We proceed to define the set of feasible *instantaneous* transmission rates that can be achieved through all possible rate and power control actions. Note that the set of feasible rates depends on the destination structures as well, that is whether the destinations in the network have SPR or MPR capabilities. Let $\mathcal{R}^{\mathrm{SPR}}(\mathbf{g})$ and $\mathcal{R}^{\mathrm{MPR}}(\mathbf{g})$ be the feasible instantaneous rate regions when the channel state is $\mathbf{G}(t) = \mathbf{g}$, $\mathbf{g} \in \mathcal{G}$ under SPR and MPR respectively. Since under SPR a transmission cannot be successful if more than one source transmits to the same destination, we first identify subsets of the sources with non-overlapping destination sets. Towards this end, we define a valid *activation vector* $\mathbf{c}$ to be a binary T-element vector that takes values in $\{0,1\}^T$. All the non-zero entries of a valid activation vector correspond to sources with non-overlapping destination sets that can be activated successfully under some rate and power allocation. In other words, for any two elements $c(n), c(n')$ of a valid activation vector $\mathbf{c}$ with $c(n) = c(n') = 1$ it must be true that $\mathcal{D}(n) \cap \mathcal{D}(n') = \emptyset$. We further define the *constraint set* $\mathcal{C}$ to be the set containing *all* such activation vectors. Based on the above, $\mathcal{R}^{\mathrm{SPR}}(\mathbf{g})$ is defined as follows:

$$\mathcal{R}^{\mathrm{SPR}}(\mathbf{g}) = \mathrm{co}\Bigg\{\mathbf{r} = (r_n,\, n \in \mathcal{T}) : \exists \mathbf{P} \in \mathcal{P},\, \exists \mathbf{c} = (c(n),\, n \in \mathcal{T}) \in \mathcal{C},$$
$$\text{such that } \frac{c(n)P_n(t)G_{(n,d)}(t)}{N_0 + \sum_{n' \in \mathcal{T}, n' \neq n} c(n')P_{n'}(t)G_{(n',d)}(t)} \geq \gamma_d(r_n)$$
$$\forall n \in \mathcal{T}, d \in \mathcal{D}(n)\Bigg\}, \tag{3.2}$$

where co(·) denotes the convex hull of a set. Hence, the set $\mathcal{R}^{\mathrm{SPR}}(\mathbf{g})$ is the set obtained by time-sharing of feasible rates, achieved by some power vector $\mathbf{P} \in \mathcal{P}$ such that concurrent transmission from two or more sources to a common destination is prohibited. Similarly, $\mathcal{R}^{\mathrm{MPR}}(\mathbf{g})$ is given by

$$\mathcal{R}^{\mathrm{MPR}}(\mathbf{g}) = \mathrm{co}\Bigg\{\mathbf{r} = (r_n,\, n \in \mathcal{T}) :\ \exists \mathbf{P} \in \mathcal{P}, \text{ such that}$$
$$\frac{P_n(t)G_{(n,d)}(t)}{N_0 + \sum_{n' \in \mathcal{T}, n' \neq n} P_{n'}(t)G_{(n',d)}(t)} \geq \gamma_d(r_n),$$
$$\forall n \in \mathcal{T}, d \in \mathcal{D}(n)\Bigg\}. \tag{3.3}$$

3.1.2 Problem Formulation

Our objective is to maximize the sum of utilities of all destinations where the utility can be a general function of the long-term average throughput. We assume utility functions $U(\cdot)$ that are strictly concave, increasing, and continuously differentiable functions of the received user rates. The utility maximization problem under consideration is

$$\max_{\bar{\mathbf{r}} \in \bar{\mathcal{R}}} \sum_{n \in \mathcal{T}} |\mathcal{D}(n)| U(\bar{r}(n)), \tag{3.4}$$

where $U(\bar{r}(n))$ is the utility of any destination $d \in \mathcal{D}(n)$ that receives from the nth source. By multiplying this utility by the number of destinations in the multicast group, we obtain the overall utility of all destinations in that group. Furthermore, the rate region $\bar{\mathcal{R}}$ in (3.4) is the long-term average rate region obtained through any stationary and ergodic policy that operates by choosing at each slot a power vector $\mathbf{P} \in \mathcal{P}$ and a rate vector that belongs either in (3.2) if the receivers have SPR capabilities or in (3.3) in the case of MPR. The rate region $\bar{\mathcal{R}}$ will be formally defined later in this section.

As an example, a utility function that satisfies these properties is the utility of α-fairness. Specifically, a utility function is called α-fair [43] if

$$U^{\alpha}(r) = \begin{cases} \log(r), & \text{if } \alpha = 1 \\ (1-\alpha)^{-1} r^{1-\alpha}, & \text{otherwise.} \end{cases} \tag{3.5}$$

The parameter α denotes the amount of "fairness" the utility function provides to the users. For $\alpha = 0$, it yields the objective of sum throughput maximization and the optimal policy activates the source with the best channel conditions (highest instantaneous rate) to its destination(s) at all times [54, 73]. For $\alpha = 1$, the function $U^{\alpha}(r)$ corresponds to the utility of *proportional fairness*, and as α increases to infinity it leads to *max–min* fairness.

Next, we formally define the long-term average rate region $\bar{\mathcal{R}}$ in (3.4). Note that this region depends on the instantaneous feasible rate region and thus it is dependent on the receiver capabilities. In the case of SPR capability, we will denote the long-term average rate region by $\bar{\mathcal{R}}^{\mathrm{SPR}}$ and it will be given by

$$\bar{\mathcal{R}}^{\mathrm{SPR}} = \Big\{ \bar{\mathbf{r}} = (\bar{r}_n, n \in \mathcal{T}) : \exists \mathbf{r}^{\mathbf{g}} = (r_n^{\mathbf{g}}, n \in \mathcal{T}) \in \mathcal{R}^{\mathrm{SPR}}(\mathbf{g}), \forall \mathbf{g} \in \mathcal{G}, \\ \text{and } \exists \mathbf{P} \in \mathcal{P} \text{ such that } \bar{r}_n = \int_{\mathcal{G}} r_n^{\mathbf{g}} f_{\mathbf{G}}(\mathbf{g}) d\mathbf{g} \Big\}. \tag{3.6}$$

This region corresponds to the long-term average rate region of throughput rates that are achievable. Note that due to the assumption of ergodicity of the channel process, the long-term average rates converge to the corresponding ensemble averages. The corresponding average rate region under MPR capabilities, $\bar{\mathcal{R}}^{\mathrm{MPR}}$, is defined similarly by replacing in (3.6) $\mathcal{R}_{\mathbf{g}}^{\mathrm{SPR}}(t)$ with $\mathcal{R}_{\mathbf{g}}^{\mathrm{MPR}}(t)$. To keep notation simple, in the rest of this section we will write $\mathcal{R}(\mathbf{g})$ to refer to the instantaneous feasible rate region $\mathcal{R}^{\mathrm{MPR}}(\mathbf{g})$ or $\mathcal{R}^{\mathrm{SPR}}(\mathbf{g})$, depending on the receiver capabilities, and, we will write $\bar{\mathcal{R}}$ for the corresponding average rate region.

Note that the problem described in (3.4) is a convex optimization problem. One approach to solve this problem is to use interior point methods to obtain the optimal vector of average rates and then to map these average rates to a sequence of instantaneous transmission rate selections (and corresponding powers) over time. However, due to the large number of variables involved in the optimization, finding these instantaneous rates is prohibitively difficult. Furthermore, the complexity of this reverse process is exacerbated by the fact that at any given slot there may be more than one such instantaneous rate selection that yields a given long-term average rate vector. Here, we follow a different approach by introducing an on-line, gradient-based solution that at every time slot selects the instantaneous rates so that the long-term average rate is the maximizer of (3.4).

3.2 An Optimal Rate and Power Allocation Control Policy

In this section, we specify an optimal, centralized policy $\boldsymbol{\pi}^\star$, which takes rate and power control decisions at the beginning of each time slot so that the achievable long-term average rate is a maximizer in (3.4). We will often write $\boldsymbol{\pi}$ to indicate a combined rate and power vector.

Let $r_n^{\pi}(t)$ denote the transmission rate of source n under policy $\boldsymbol{\pi}$ at time slot t and let $\mathbf{r}^{\pi}(t)$ denote the T-dimensional vector of rates $\mathbf{r}^{\pi}(t) = (r_n^{\pi}(t), n \in \mathcal{T})$. Also, define by $\theta_n^{\pi}(t)$ the time-average rate of source n under a policy $\boldsymbol{\pi}$ *until* time slot t. We will refer to this time-average rate as the *effective rate* of source n at time slot t (as opposed to the *instantaneous rate* at time slot t). This effective rate coincides with the rate that each destination $d \in \mathcal{D}(n)$ receives *up* to time slot t. Specifically,

$$\theta_n^{\pi}(t) = \frac{1}{t}\sum_{\tau=1}^{t} r_n^{\pi}(\tau). \tag{3.7}$$

We also define the vector of effective rates of all sources up to time slot t by $\boldsymbol{\theta}^{\pi}(t) = (\theta_n^{\pi}(t), n \in \mathcal{T})$. Assume that the effective rates at time slot $t-1$ are given by $\boldsymbol{\theta} = \boldsymbol{\theta}^{\pi^\star}(t-1)$. Then, the *power and rate control* policy $\boldsymbol{\pi}^\star$ is defined as

$$[\mathbf{P}^{\boldsymbol{\pi}^\star,\ \boldsymbol{\theta},\mathbf{g}}(t), \mathbf{R}^{\boldsymbol{\pi}^\star,\ \boldsymbol{\theta},\mathbf{g}}(t)] = \left\{ \arg\min_{\mathbf{P}\in\mathcal{P},\mathbf{r}\in\mathcal{B}^{\boldsymbol{\theta}}(\mathbf{G}(t))} \sum_{n\in\mathcal{T}} P_n \right\}, \tag{3.8}$$

where $\mathcal{B}^{\boldsymbol{\theta}}(\mathbf{G}(t))$ is given by

$$\mathcal{B}^{\boldsymbol{\theta}}(\mathbf{G}(t)) = \left\{ \tilde{\mathbf{r}} : \tilde{\mathbf{r}} = \arg\max_{\mathbf{r}\in\mathcal{R}(\mathbf{G}(t))} \sum_{n\in\mathcal{T}} |\mathcal{D}(n)| r_n \frac{\partial U(\theta_n)}{\partial \theta_n} \right\}. \tag{3.9}$$

Note that the set $\mathcal{B}^{\boldsymbol{\theta}}(\mathbf{G}(t))$ may not be well defined when the gradient of the utility function is not finite. Such an example is the utility of proportional fairness where for θ_n close to 0 the gradient is ill-defined. However, this problem can be overcome by adding arbitrarily small initialization constants to the argument of the gradient of the utility, as was done in [38, 46]. The optimality of $\boldsymbol{\pi}^\star$ defined in (3.8) and (3.9), with respect to the utility maximization problem of (3.4), was established in [46]. The proof in [46] followed stochastic approximation arguments.

3.3 Simulation Results

In this section, we analyze the performance of the optimal policy of Section 3.2 through a set of simulations both for unicast and multicast traffic. Throughout this section, we consider the utility of proportional fairness. We observe the performance of our policy both in the presence and absence of channel fading. As we will see later in this section, unicast traffic significantly benefits from the presence of fading. These benefits are reduced in the case of multicast.

Throughout our simulations we consider a single-hop, wireless network with three sources and three destinations. The duration of a time slot is assumed to be equal to 1 s. We only consider rate control, and assume that each source n at every time slot t can either remain silent or transmit at a maximum power level P_n, $n = 1,2,3$. Specifically, we assume that the transmission powers satisfy $P_1 = 6.0 \times 10^{-5}$ W, $P_2 = 3.0 \times 10^{-5}$ W, and $P_3 = 2.0 \times 10^{-5}$ W. Further, the power of the Additive White Gaussian Noise (AWGN) is assumed to be $N_0 = 3.34 \times 10^{-6}$ W at all destinations.

In our model we consider quasi-static, Rayleigh fading. Let the received signal power under path loss and shadowing between source n and destination d at time slot t be denoted by $P^r_{(n,d)}(t)$. Moreover, let the average received power be $\bar{P}^r_{(n,d)}$. Then, under Rayleigh fading the received signal power $P^r_{(n,d)}(t)$ is exponentially distributed with mean $\bar{P}^r_{(n,d)}$ (see e.g., Section 3 in [25]).

Let us define the average path loss matrix $\bar{\mathbf{G}} = (\bar{G}_{(n,d)}, n, d = 1,2,3)$, which is obtained by averaging the path losses over all different states of the channel (fading) process $\{\mathbf{G}(t)\}_{t=0}^{\infty}$. With the common assumption that the shadowing process varies according to a zero mean Gaussian random variable it follows that the channel coefficients $G_{(n,d)}(t)$ satisfy $G_{(n,d)}(t) = P^r_{(n,d)}(t)/P_n$ at time t and furthermore the average path loss is equal to $\bar{G}_{(n,d)} = \bar{P}^r_{(n,d)}/P_n$. We study the performance of our policy under two scenaria, namely under (i) pure path loss and (ii) Rayleigh fading. For the comparison to be meaningful, we assume that the matrix of path losses under scenario (i) is given by the matrix $\bar{\mathbf{G}}$, which is the mean of the Rayleigh fading of scenario (ii). The matrix $\bar{\mathbf{G}}$ is parameterized as follows:

$$\bar{\mathbf{G}} = \begin{bmatrix} 0.9 & \beta \times 0.9 & \beta \times 0.9 \\ \beta \times 0.9 & 0.9 & \beta \times 0.9 \\ \beta \times 0.9 & \beta \times 0.9 & 0.9 \end{bmatrix}, \tag{3.10}$$

where $\beta \in [0,1]$ multiplies the cross channel path losses. For the case of unicast traffic, we dub the parameter β as the *interference coefficient* while we call it *cross-link coefficient* for the case of multicast. In the case of unicast traffic, this parameter reflects the level of interference in the network. For example, when $\beta = 0$, the channels between the three source/destination pairs can be seen as three parallel channels that can operate simultaneously without causing any interference to each other. On the other extreme, when $\beta = 1$, the path losses at the direct channels between every source and destination are equal to the path losses over the cross channels, and therefore the level of interference at every destination is very high. In the case of multicast traffic the parameter β gives the quality of the cross links and has an effect not only on the interference, but also on the transmission rate. Throughout the section, we obtain the data rate $r(\cdot)$ at the destinations through the single-user Shannon formula, i.e., $r(\text{SINR}) = \log_2(1 + \text{SINR})$, by assuming unit bandwidth. A more precise and practically meaningful option would be to use the relationship between rate and threshold that corresponds to a specific modulation scheme as it was considered in [48]. However, these relationships are given by much more complicated formulas. The qualitative comparisons using the Shannon formula are not affected significantly but they are indeed only approximate.

3.3.1 Case I — Unicast Sessions

In this section, we consider the special case of unicast sessions, i.e., $\mathcal{D}(1) = \{1\}$, $\mathcal{D}(2) = \{2\}$, and $\mathcal{D}(3) = \{3\}$.

In Figure 3.2, we present the convergence of effective rates of each destination when we employ the optimal policy presented in Section 3.2 under a fading and a non-fading channel model. The interference coefficient is set to $\beta = 0.2$. From Figure 3.2 it follows that the

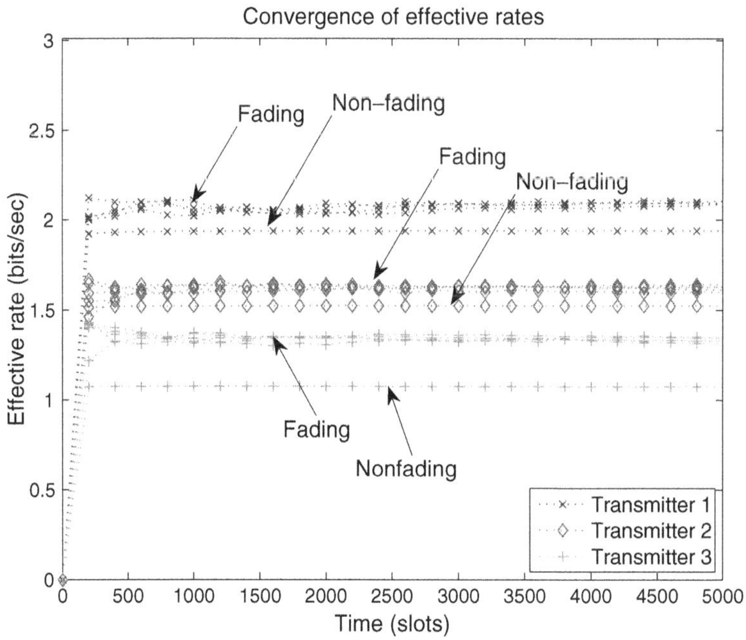

Fig. 3.2 Convergence of the utility optimal policy $\pi^\star$ for unicast traffic.

effective rate of each destination quickly converges to its corresponding proportionally fair rate. We also observe that, naturally, the effective rates are increasing functions of the transmission powers, and thus destination one has a higher rate than destination two, and the latter has a higher rate than destination three.

From the figure we can also observe that the effective rate of each destination is higher under fading than in the absence of fading. This demonstrates the opportunistic nature of the optimal policy. If a source sees a bad channel at the current time slot, it will not activate this source in general, since with *high probability* in the future its channel conditions are expected to improve. Moreover, at the current time slot with high probability some other source having a better channel will be activated instead.

We proceed to compare the proposed optimal policy (i) when it can take all possible rate control actions in the set $\mathcal{R}(\mathbf{G}(t))$ and (ii) when it is restricted to only use TDMA scheduling and rate control actions, i.e., it can only activate a single source at any given time at the maximum achievable corresponding rate. We refer to the former as the "optimal rate control policy" and to the latter as the "TDMA scheduling policy". Note that since the transmission powers take values from the set $\{0, P_n\}$ for every source n, the set $\mathcal{R}(\mathbf{G}(t))$ for every time slot t contains seven rate vectors obtained by finding all possible subsets of sources and assigning them the maximum transmission rates such that the SINR criterion is jointly satisfied at their destinations. The comparison of the two policies is performed under various interference levels.

In Figure 3.3 the proportionally fair effective rates at each destination are plotted as a function of different values of the interference coefficient β for the optimal rate control policy and for the TDMA scheduling policy in the absence of fading. We observe that when the interference levels are relatively low, the optimal rate control policy achieves higher rates for every source and destination pair, compared to the TDMA scheduling policy. We also observe that the two policies have comparable performance under higher interference levels. This result is natural since the proposed policy exploits the potential benefits of concurrent transmissions when the interference is relatively low and effectively operates as TDMA scheduling when the interference is

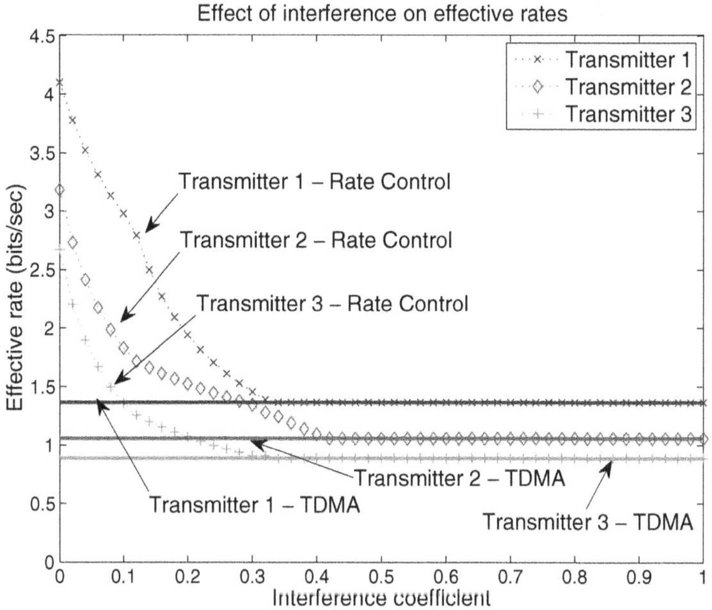

Fig. 3.3 Proportionally fair rates as a function of the interference coefficient β in the absence of fading under the optimal rate control policy and under the TDMA scheduling policy for the unicast traffic case.

relatively high. A similar pattern is observed in Figure 3.4, where fading is considered.

3.3.2 Case II — Multicast Sessions

In this section, we consider the case where there are three sessions, one broadcast, one multicast, and one unicast. Specifically, $\mathcal{D}(1) = \{1,2,3\}$, $\mathcal{D}(2) = \{1,2\}$, and $\mathcal{D}(3) = \{3\}$. We assume MPR capabilities at the destinations, e.g., both sessions 1 and 2 can be activated simultaneously as long as the SINRs for each received transmission at every destination exceed the appropriate thresholds. In Figure 3.5, we present the convergence of the effective rates of each destination achieved under the optimal policy by setting the cross-link coefficient equal to $\beta = 0.2$.

Again in this figure we observe that the effective rate of each multicast session converges quickly to its respective proportionally fair rate. We further observe that unlike the unicast cases, here it is no longer

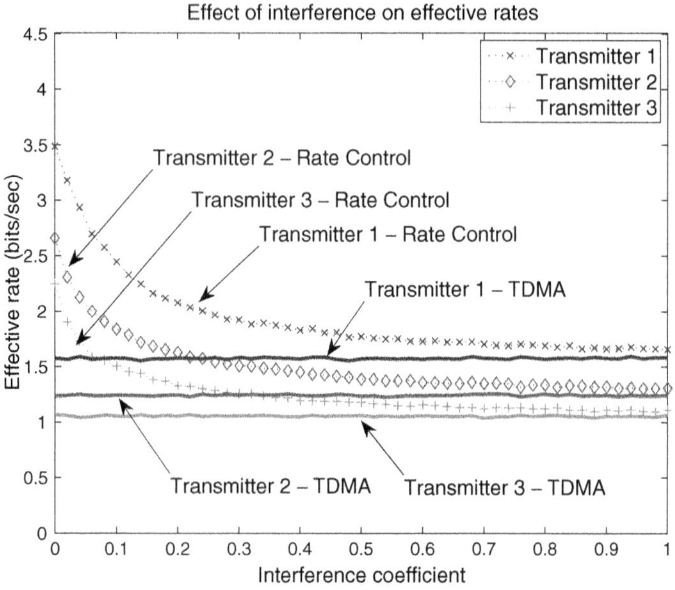

Fig. 3.4 Proportional fair rates as a function of the interference coefficient β in the presence of Rayleigh fading under the optimal rate control policy and under the TDMA scheduling policy for the unicast traffic case.

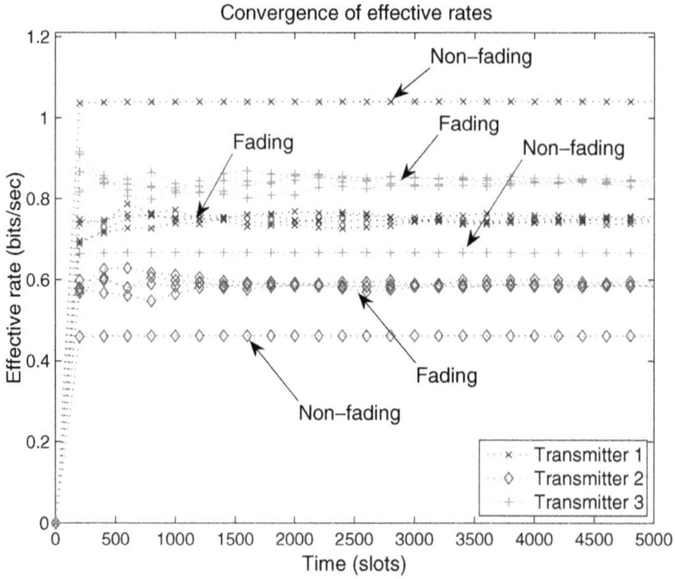

Fig. 3.5 Convergence of the utility optimal policy $\boldsymbol{\pi}^\star$ for multicast traffic.

true that the rates under fading are always better than in the absence of fading. The reason behind this fact is that, now, a transmission involves multiple links and the multicast rate is constrained by the link with the worst channel due to the single rate multicast assumption. Hence, for source 1 to effectively observe a "good" channel, all the three channels to which it broadcasts have to be good simultaneously. Clearly, the probability of occurrence of this event decreases as the number of destinations of a multicast session increases. Therefore, the average received multicast rate of the broadcast session is naturally worse under fading. On the other hand, the average received rate under multicast session 2 and under the unicast from source 3 is still better under fading due to the opportunistic nature of the optimal policy and the reduced effect of the "worst" channel in each group.

In Figures 3.6 and 3.7 the proportionally fair rates of each multicast session are plotted as functions of the cross-link coefficient β under

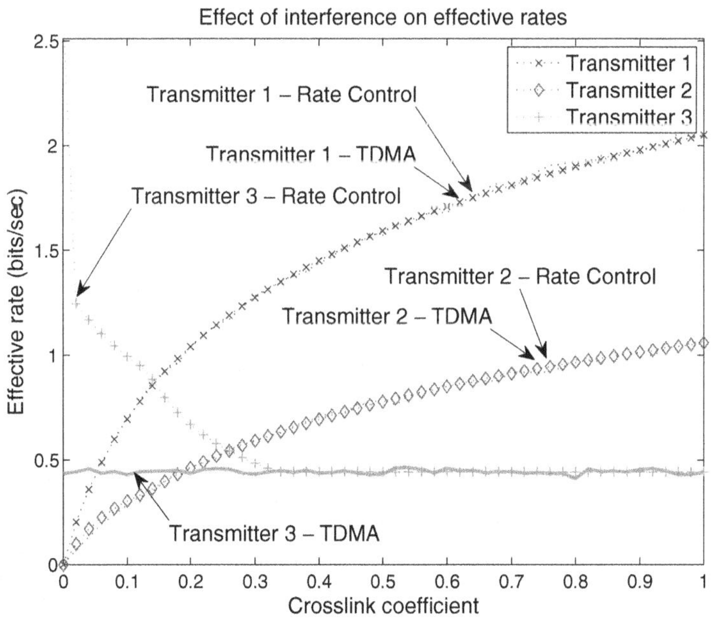

Fig. 3.6 Proportionally fair rates as a function of the cross-link coefficient β in the absence of fading under the optimal rate control policy and under the TDMA scheduling policy for the multicast traffic case.

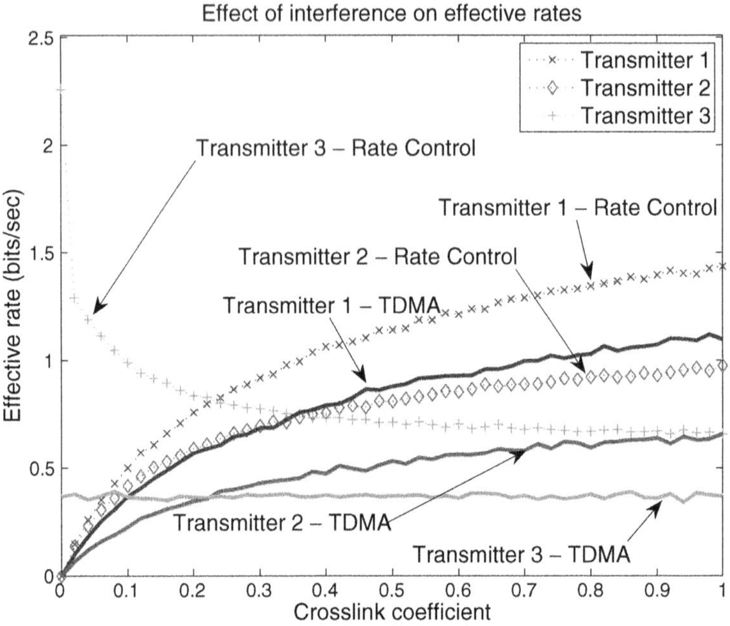

Fig. 3.7 Proportionally fair rates as a function of the cross-link coefficient β in the presence of Rayleigh fading under the optimal rate control policy and under the TDMA scheduling policy for the multicast traffic case.

non-fading and under Rayleigh fading. As in the unicast case, we plot both the optimal rate control policy, as well as the TDMA scheduling policy. From Figure 3.6 we observe that the only session that enjoys a higher rate under the optimal rate control policy is the unicast session. For the broadcast transmission from source 1 and the multicast transmission from source 2, no benefits are observed under concurrent operation of the sources, even for small values of the parameter β. The reason for this is the multi-packet reception capability of the destinations and that the quality of the direct links is fixed and equal to 0.9. For example, in the case of broadcasting from source 1, the high quality of the direct link will not only potentially increase the rate of the broadcast session using this link, but will also cause high interference to the multicast session from source 2 using destination 1. The same observation is true for the multicast session from source 2. Therefore, regardless of the value of the cross-link coefficient, β, rate control does not provide any additional gains in terms of rate compared to TDMA for the multicast and broadcast cases.

However, the above conclusions are valid only for non-fading channels. As shown in Figure 3.7 the gains achieved by allowing more sessions to operate concurrently reduces the TDMA performance when fading is present.

4

Reduction of Scheduling Complexity

In this section, we revisit the problem of joint scheduling and rate control for static, single-hop wireless networks. We consider both the performance objectives of utility maximization and minimum-length scheduling when each source can either transmit at its maximum transmission power or remain silent. In such a setting and when the transmission rates of the sources can be controlled, there exist $2^T - 1$ possible ways to activate T sources. Clearly, as we have discussed already in Section 2, scheduling and rate control is a problem of combinatorial complexity. There are several approaches to reduce the complexity of such problems. One approach is to obtain heuristics that approximate the optimal solution at the benefit of a decreased complexity. An example along these lines was followed in [33] and [34] where the authors solve the minimum-length scheduling problem by formulating it as a linear program and by using the technique of column generation to obtain a solution with a better running time. In this section, we are going to focus on a different approach. Specifically, we are going to reduce the search space of possible rate control and scheduling decisions and solve several instances of the scheduling problem "optimally", within this reduced set of actions.

We will first consider the performance objective of *sum throughput* maximization under a slotted time model. Since maximizing the sum throughput of the network can be unfair to users facing poor channel conditions, we will next consider the objective of proportional fairness which has been shown to provide a good compromise between efficiency and fairness [53]. We will formulate the problem of obtaining an optimal proportionally fair policy as a convex problem. Despite its convexity, the combinatorial complexity of this problem prohibits us from characterizing the optimal solution. Thus, next, by restricting the action space, we will analytically solve the corresponding convex problem and obtain a suboptimal, polynomial-time solution, that is the optimal proportionally fair rate allocation over the *reduced set* of rate control and scheduling decisions. Finally, we will look at the *minimum-length scheduling* problem under a continuous time model. By restricting the set of feasible rate control and scheduling actions, we will obtain a suboptimal, but polynomial time, solution to the problem. This approach on state space reduction was previously introduced in [48] under the objective or proportional fairness for unicast traffic, in [49, 50] under a more general objective of utility maximization for multicast traffic, and in [47] under the objective of minimum-length scheduling.

4.1 Model Formulation

We consider the model described in Section 3 where T sources of traffic wish to multicast to D destinations. Each source $n \in \mathcal{T}$ multicasts at a *common* rate to a set of destinations $\mathcal{D}(n) \subseteq \mathcal{D}$. As before, our formulation includes the case of unicast and broadcast traffic as special cases. Furthermore, as previously, we will assume that a destination $d \in \mathcal{D}$ can receive from more than one multicast source, i.e., for multicast sources $n, n' \in \mathcal{T}$, it is possible that $\mathcal{D}(n) \cap \mathcal{D}(n') \neq \emptyset$. In the specific case of unicast traffic for the purpose of simplifying notation we will assume that $T = D$ and that the sources and the destinations form T communication pairs. We assume that the sources of traffic can either operate at their maximum transmission power level or remain silent.

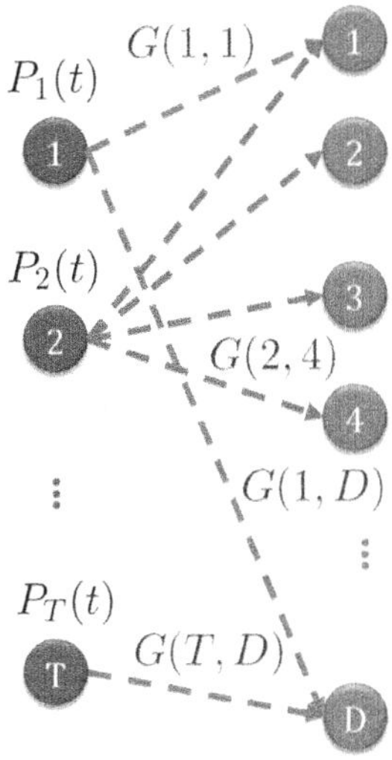

Fig. 4.1 A single-hop network of T sources of traffic and D destinations.

As before, $P_n(t)$ denotes the transmission power level of source n at time slot t. In this section, we assume that the variable $P_n(t)$ can take two possible values, namely $P_n^{\max}$ (when source n is activated) and 0 (when it remains silent). We denote by $\mathbf{P}(t)$ the T-dimensional vector of transmission powers at time slot t, i.e., $\mathbf{P}(t) = (P_n(t),\ n \in \mathcal{T})$. Moreover, we assume that the channel conditions between every source and destination in the network do not vary with time and are due to pure path loss (that is there is no fading). We denote by $G(n,d)$ the path loss between source $n \in \mathcal{T}$ and destination $d \in \mathcal{D}$. The network model we consider is depicted in Figure 4.1.

In the case of multicasting successfully the same message to a set of destinations we assume, as before, that the SINR criterion has to be satisfied at each one of them. When the transmission powers at the sources take binary values, a scheduling decision amounts to selecting a scheduling and rate control vector out of $2^T - 1$ possible choices. In this section, we will call these $2^T - 1$ possible scheduling and rate control decisions as *actions*. Let us denote by $\mathcal{A}$ the set of all possible actions, i.e., $|\mathcal{A}| = 2^T - 1$. This corresponds to $\mathcal{A} = \mathcal{R}$ in Section 2.3. The optimal rates at which the multicast sources should operate at any given time depend on the adopted performance objective and on the link channel conditions.

We denote by r_n^j the *instantaneous* rate at which source $n \in \mathcal{T}$ transmits to all of the destinations $\mathcal{D}(n)$ in its multicast session, under Action $j \in \mathcal{A}$. Since we consider single-rate multicast, the rate of source n is equal to the rate of every destination d in its multicast session, i.e., $d \in \mathcal{D}(n)$. Thus, we can characterize the rate of each destination $d \in \mathcal{D}(n)$ through the transmission rate of its corresponding source. Let $\mathbf{f} = (f_0, \ldots, f_{|\mathcal{A}|-1})$ denote a probability distribution over the set of all possible rate control and scheduling actions in $\mathcal{A}$. That is, we randomize the policy decision so that in every slot Action j is taken with probability f_j. This formulation by-passes one aspect of combinatorial complexity that arises when we associate each action in a deterministic way with each slot. We assume that such probability distribution exists, e.g., by requiring ergodicity on the action selection. Since a source is not activated at the same rate in every slot, we define the *effective rate* $r_n(\mathbf{f})$ of source $n \in \mathcal{T}$ to be the average rate over the action distribution $\mathbf{f}$, i.e.,

$$r_n(\mathbf{f}) = \sum_{j \in \mathcal{A}} r_n^j f_j.$$

Here again, we would like to note that we define multicast throughput to be the *overall* traffic that reaches all the destinations of a multicast session, i.e., the throughput of source n is equal to

$$|\mathcal{D}(n)| \sum_{j \in \mathcal{A}} r_n^j f_j.$$

Thus, for any two multicast sources that operate at equal rates, the source that has a higher number of destinations is assumed to contribute more in terms of throughput. In other words, our criterion is the *received throughput* which reflects the number of destinations in the multicast session. Note that it is also possible to require that all destinations must receive a message in order to count it against throughput (which we do here), or, alternatively, to simply count whatever messages are successfully received without insisting that all destinations receive every message.

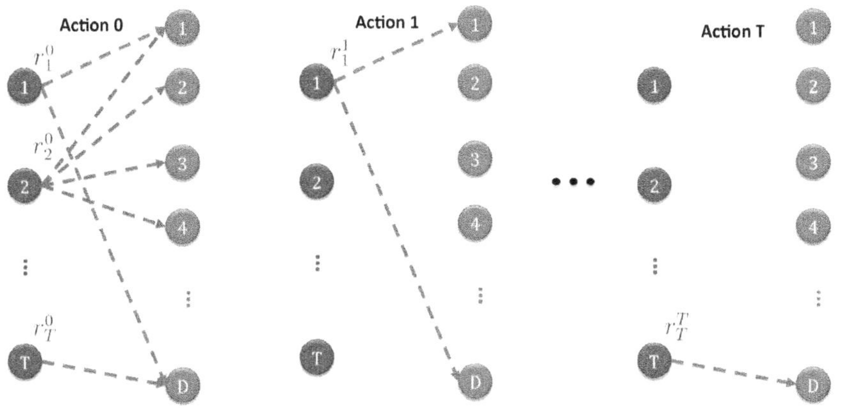

Fig. 4.2 The $T+1$ possible actions obtained by either scheduling T sources "one-at-a-time" or "all-at-once". The rate of source n under Action j is denoted by r_n^j.

4.1.1 State Space Reduction

Instead of considering all the $2^T - 1$ possible scheduling and rate control actions at which the multicast sources can be scheduled, in this section we consider only a subset of those given by the following two simple schemes. Specifically, these actions include (i) the simultaneous activation of *all* T multicast sources operating successfully and at instantaneous rates that ensure all SINR threshold inequalities are satisfied (we call this operation "all-at-once" or "Action 0") and (ii) the individual activation of each source separately, as in TDMA (we call this operation "one-at-a-time" or "Action n" when source n is activated). Clearly, under Action n the instantaneous rate is the maximum possible rate that permits the SINR (which in this case is merely the Signal to Noise Ratio) for the given transmission power to exceed the corresponding threshold at each destination $d \in \mathcal{D}(n)$. However, since this source is not transmitting at all times its rate on the average can be much lower. The above two modes of operation yield a total of $T+1$ actions, as shown in Figure 4.2.

4.2 Total Throughput Maximization

In this section, we obtain a scheduling and rate control policy that maximizes the total (sum) received throughput of the network. This maximization problem can be posed as follows:

Problem I :

$$\max_{\mathbf{f}} \sum_{n \in \mathcal{T}} |\mathcal{D}(n)| r_n(\mathbf{f}) \tag{4.1}$$

s.t.

$$f_j \geq 0, \quad j \in \mathcal{A}, \tag{4.2}$$

$$\sum_{j \in \mathcal{A}} f_j = 1. \tag{4.3}$$

Note that, if throughput were measured with respect to the *transmitted* rate, all the weights $|\mathcal{D}(n)|$ would have been equal to one in the maximization problem described in (4.1)–(4.3). Consider also the closely associated surrogate problem, defined as

Problem II :

$$\max_{j \in \mathcal{A}} \sum_{n \in \mathcal{T}} |\mathcal{D}(n)| r_n^j. \tag{4.4}$$

The following proposition shows how these two problems relate to each other.

Proposition 4.1. Let $\mathcal{A}^\star \subseteq \mathcal{A}$ denote the set of actions solving Problem II defined in (4.4). The optimal probability assignment solving Problem I defined in (4.1)–(4.3) satisfies $\sum_{j \in \mathcal{A}^\star} f_j = 1$.

The proof of Proposition 4.1 appears in [50]. This proposition says that the set of scheduling actions that are assigned a non-zero probability are those that maximize the total received rate in (4.4). If there is a unique such scheduling action it will be used always with probability one. If there exist more than one such action, then they can be assigned a probability distribution arbitrarily. In fact, it suffices that only one of them is used at all times (i.e., a single action is chosen and is assigned a probability equal to one).

Note that although the problem in (4.4) is a linear program, characterizing the set $\mathcal{A}^\star$ is a difficult task. This is because finding the maximizer(s) action(s) involves evaluation of the sum throughput achieved by each element in the set $\mathcal{A}$, which is a set that grows exponentially in the number of multicast sources.

Consequently, instead of looking at all possible actions, we are going to restrict attention to the two modes of operation described in Section 4.1.1. Restricting attention to the above two modes of operation is somewhat natural since it permits comparison between two extreme cases, namely the cases of operation "all-at-once" and "one-at-a-time". Although under Action 0 all sources operate concurrently, their individual rates will likely be low so that the SINR criterion is satisfied under the heavy interference that occurs when they all transmit at the fixed power levels.[1] On the other hand, although under Action n the instantaneous rate of the nth source will likely be much higher (than the corresponding rate under concurrent operation), its rate on the average may be much lower due to the effect of time sharing. Although this represents a severe restriction of the action space, it is expected to provide an insight into the trade-off between concurrent and individual activation. The optimal solution to the restricted problem is the following:

Corollary 4.2. The optimal solution to Problem I defined in (4.1)–(4.3) when the set $\mathcal{A}$ is restricted to contain only the actions of operation "all-at-once" and "one-at-a-time", i.e., $\mathcal{A} = \{0, 1, \ldots, T\}$ is the following: Find the multicast source that maximizes the overall instantaneous received rate under operation "one-at-a-time", i.e.,

$$n^\star = \arg\max_{n \in \mathcal{T}} |\mathcal{D}(n)| r_n^n. \tag{4.5}$$

If

$$|\mathcal{D}(n^\star)| r_{n^\star}^{n^\star} \geq \sum_{n \in \mathcal{T}} |\mathcal{D}(n)| r_n^0, \tag{4.6}$$

then the multicast source $n^\star$ is selected to operate at all times individually (Action $n^\star$ is chosen), otherwise all multicast sources operate concurrently (Action 0 is chosen).

Note that by restricting attention to only these two schemes the scheduling problem under the objective of sum throughput maximization is significantly simplified and becomes linear in the number of the network sources (instead of exponential). Furthermore, if larger

[1] Note that we do not consider power control here.

portions of the state space had been used the nature of the approach would remain the same. However, we would expect an improvement in the sub-optimality of the solution. At the same time, this improvement would also increase the complexity of solving this problem. Thus, by considering this simplified version of the problem we are able to build insights on how the optimal solution is affected by the current channel conditions.

It is clear that optimizing the sum throughput of the network leads to an efficient utilization of the network resources since the network sends traffic at the maximum rate that it can support. Nevertheless, it can lead to serious unfairness among the sources since the optimal action set may totally exclude sources with poor channel conditions, prohibiting them from accessing the channel. In the next section, we consider the utility of proportional fairness [30] which has been widely used as a performance metric in wireless networks.

4.3 Proportional Fairness

In this section, we focus on the objective of proportional fairness. Consider a rate vector $\mathbf{r} = (r_d, d \in \mathcal{D})$ of average feasible reception rates at each destination. The vector $\mathbf{r}$ is said to be *proportionally fair* [31] if for any other such average feasible rate vector $\mathbf{r}' = (r'_d, d \in \mathcal{D})$ it is true that

$$\sum_{d \in \mathcal{D}} \frac{r'_d - r_d}{r_d} \leq 0.$$

Recall that r_n^j is the instantaneous transmission rate of source n under Action j. We are interested in obtaining an optimal probability distribution so that the *effective* rates of each destination $d \in \mathcal{D}$ are assigned in a proportionally fair way. It was shown in [30] that the objective of proportional fairness is equivalent to maximizing the sum of the logarithms of the user rates over all the feasible rate vectors. Since the multicast transmissions we consider are single-rate multicast, it follows that for any source $n \in \mathcal{T}$ all its destinations $d \in \mathcal{D}(n)$ receive at the same rate. Thus, we can write this utility maximization problem with respect to the feasible transmission rates of the sources. This can be expressed as a convex optimization problem as follows:

$$\max_{\mathbf{f}} \sum_{n \in \mathcal{T}} |\mathcal{D}(n)| \log\left(r_n(\mathbf{f})\right) \tag{4.7}$$

s.t.

$$f_j \geq 0, \quad \forall j \in \mathcal{A}, \tag{4.8}$$

$$\sum_{j \in \mathcal{A}} f_j = 1, \tag{4.9}$$

where f_j is the probability that in a given slot Action j is chosen. Although this is a convex problem, the number of possible actions, and hence constraints, increases exponentially in the number of multicast sources. Therefore, although, when the number of sources in the network is sufficiently small, it can be solved numerically (for example, through interior-point methods [13]), in general computing the optimal solution analytically is infeasible.

Consequently, in what follows, we restrict our attention to the aforementioned schemes of operation "all-at-once" and "one-at-a-time". Instead of computing the optimal solution with respect to all exponentially many rate control and scheduling decisions, we will find the optimal solution of this restricted problem. This will significantly simplify the scheduling policy and render the problem polynomial. Again, since by reducing the set of actions from $2^T - 1$ to $T + 1$ the restriction in the action space is significant, we do not hope for optimality. Still, we expect to obtain intuition about which of these "extreme" modes of operation is preferable depending on the current channel conditions. Let $\tilde{\mathbf{f}}$ be a probability distribution over the restricted set of actions, i.e., $\tilde{\mathbf{f}} = (\tilde{f}_0, \tilde{f}_1, \ldots, \tilde{f}_T)$. Next, we find the optimal proportionally fair probability distribution over the aforementioned restricted set of actions by solving the following problem:

$$\max_{\tilde{\mathbf{f}}} \sum_{n \in \mathcal{T}} |\mathcal{D}(n)| \log(\tilde{f}_0 r_n^0 + \tilde{f}_n r_n^n) \tag{4.10}$$

s.t.

$$\tilde{f}_j \geq 0, \quad \forall j \in \{0, 1, \ldots, T\}, \tag{4.11}$$

$$\sum_{j=0}^{T} \tilde{f}_j = 1. \tag{4.12}$$

Before we characterize the optimal policy solving (4.10)–(4.12), we provide some useful definitions. Let $\mathcal{J}$ be a subset of the set $\mathcal{T}$, such that for every $j \in \mathcal{J}$ it is true that $\tilde{f}_j > 0$. Also, let the complement $\mathcal{J}^c = \mathcal{T} \setminus \mathcal{J}$ of the set $\mathcal{J}$ be a set such that for every $i \in \mathcal{J}^c$ it follows that $\tilde{f}_i = 0$.

Proposition 4.3. Let $\tilde{\mathbf{f}}^\star = (\tilde{f}_0^\star, \ldots, \tilde{f}_T^\star)$ denote the solution to (4.10)–(4.12) above. Then we have:

(1) If

$$\sum_{n \in \mathcal{T}} \frac{r_n^0}{r_n^n} \leq 1,$$

each multicast source $n \in \mathcal{T}$ is scheduled to transmit individually with probability

$$\tilde{f}_n^\star = \frac{|\mathcal{D}(n)|}{\sum_{j \in \mathcal{T}} |\mathcal{D}(j)|}, \quad \forall n \in \mathcal{T},$$

and the probability of concurrent operation satisfies $\tilde{f}_0^\star = 0$.

(2) If

$$\sum_{n \in \mathcal{T}} \frac{r_n^0}{r_n^n} > 1,$$

the optimal policy is of a threshold type with threshold $R(\mathcal{J})$ given by

$$R(\mathcal{J}) = \frac{1 - \sum_{j \in \mathcal{J}} r_j^0 / r_j^j}{\sum_{m \in \mathcal{J}^c} |D(m)|}. \tag{4.13}$$

Specifically:

(a) A multicast source $j \in \mathcal{T}$ is scheduled to transmit individually with probability $\tilde{f}_j^\star > 0$ (i.e., j is activated individually and belongs to $\mathcal{J}$) given by

$$\tilde{f}_j^\star = \frac{|\mathcal{D}(j)| - \sum_{i \in \mathcal{J}^c} |\mathcal{D}(i)| \frac{r_j^0 / r_j^j}{1 - \sum_{j \in \mathcal{J}} r_j^0 / r_j^j}}{\sum_{n \in \mathcal{T}} |\mathcal{D}(n)|}, \tag{4.14}$$

if and only if

$$\frac{r_j^0}{|\mathcal{D}(j)|r_j^j} < R(\mathcal{J}). \tag{4.15}$$

(b) All sources operate concurrently with probability $\tilde{f}_0^\star$ given by

$$\tilde{f}_0^\star = \frac{\sum_{m \in \mathcal{J}^c} |\mathcal{D}(m)|}{\left(\sum_{n\in\mathcal{T}} |\mathcal{D}(n)|\right)\left(1 - \sum_{j\in\mathcal{J}} r_j^0/r_j^j\right)}. \tag{4.16}$$

The proof of Proposition 4.3 appears in [50]. The quantity $\sum_{n\in\mathcal{T}} \frac{r_n^0}{r_n^n}$ determines, in a sense, the relative degree of interference in the network. Clearly, for any source the instantaneous rate under concurrent operation is no greater than its corresponding rate under individual transmission. However, if it is also true that $\sum_{n\in\mathcal{T}} \frac{r_n^0}{r_n^n} \leq 1$, then the sources interfere among themselves significantly, so that their corresponding rates under concurrent operation are much lower than the corresponding rates under individual operation. Hence, when $\sum_{n\in\mathcal{T}} \frac{r_n^0}{r_n^n} \leq 1$, the optimal policy would never activate all sources concurrently ($\tilde{f}_0^\star = 0$); instead the optimal scheduling and rate control solution is to activate a single source at a time, as in a TDMA scheme. On the other hand, if $\sum_{n\in\mathcal{T}} \frac{r_n^0}{r_n^n} > 1$, the interference among the sources when they concurrently transmit is not so severe, and hence, the individual rates under concurrent operation result in levels that are "comparable" to those achieved under individual operation. Thus, the optimal policy assigns a positive probability to Action 0 ($\tilde{f}_0^\star > 0$).

Proposition 4.3 characterizes the optimal solution based on the threshold function $R(\mathcal{J})$ which itself is a function of the set $\mathcal{J}$. Hence, in order to completely characterize the optimal policy we need to characterize the composition of $\mathcal{J}$. Note that since the optimal policy is of threshold type, the cardinality $|\mathcal{J}|$ of the "individually activated" set suffices to completely determine the set $\mathcal{J}$ itself, provided we label the sources appropriately. To simplify the notation in the sequel we will write $R(j)$ to denote $\{R(\mathcal{J})\colon |\mathcal{J}| = j\}$.

Let us reorder the multicast sessions with respect to their corresponding values of the ratios $r_j^0 / \left(|\mathcal{D}(j)| r_j^j\right)$, $j \in \mathcal{T}$ in increasing order, i.e.,

$$\frac{\tilde{r}_1^0}{|\tilde{\mathcal{D}}(1)|\tilde{r}_1^1} \leq \frac{\tilde{r}_2^0}{|\tilde{\mathcal{D}}(2)|\tilde{r}_2^2} \leq \cdots \leq \frac{\tilde{r}_T^0}{|\tilde{\mathcal{D}}(T)|\tilde{r}_T^T}, \tag{4.17}$$

where the rates $\tilde{r}_j^0$, $\tilde{r}_j^j$, and the set of destinations $\tilde{\mathcal{D}}(j)$ denote the quantities r_j^0, r_j^j, and $\mathcal{D}(j)$ respectively of the jth source under the new ordering. From now on, unless otherwise stated, the source j is the jth source under this new ordering. We will make use of the following property of the threshold function $R(j)$ to obtain the cardinality of the set $\mathcal{J}$.

Lemma 4.4. Under the ordering of (4.17), the threshold function $R(j)$ defined in Proposition 4.3 satisfies the following:

$$R(j-1) \leq R(j), \quad \text{if and only if } \; j \in \mathcal{J}.$$

The proof of Lemma 4.4 appears in [50]. From Lemma 4.4 it follows that $R(j)$ is increasing for all $j \in \mathcal{J}$ and decreasing for all $j \in \mathcal{J}^{\mathrm{c}}$. Using this fact, the following result follows directly using the definition of $R(\mathcal{J})$.

Proposition 4.5. The cardinality of the set $\mathcal{J}$ under the optimal policy specified in Proposition 4.3 is given by

$$|\mathcal{J}| = \operatorname*{arg\,max}_{\ell \in \{0,1,\ldots,T\}} \frac{1 - \sum_{\{j: j \geq 1, j \leq \ell\}} \tilde{r}_j^0 / \tilde{r}_j^j}{\sum_{\{m: m \geq \ell+1, m \leq T\}} |\tilde{\mathcal{D}}(m)|}. \tag{4.18}$$

From Propositions 4.3 and 4.5 it follows that the set $\mathcal{J}$ contains the $|\mathcal{J}|$ sources with the lowest values of the ratios $r_j^0/(|\mathcal{D}(j)| r_j^j)$ for $j \in \mathcal{T}$. Hence, in the optimal solution the sources that are selected to be activated individually are the most "disadvantaged" multicast sources, i.e., those that either (i) can only achieve very low rates under concurrent operation compared to individual operation or (ii) those that multicast to a large number of destinations.

4.3.1 A Special Case: Unicast Traffic

Consider a single-hop network of T source and destination pairs, where each source sends unicast traffic to its corresponding destination. Note that under the restricted set of actions the optimal proportionally fair probability distribution for the unicast case follows directly from our formulation by simply setting the cardinality of the set $\mathcal{D}(n)$ for every source $n \in \mathcal{T}$ equal to one, i.e., $|\mathcal{D}(n)| = 1$. Thus, the solution for the case of unicast traffic is provided next.

Corollary 4.6. Let $\tilde{\mathbf{f}}^{\mathrm{u}^\star} = (\tilde{f}_0^{\mathrm{u}^\star}, \ldots, \tilde{f}_T^{\mathrm{u}^\star})$ be the optimal proportionally fair probability distribution for unicast traffic. Then we have:

(1) If

$$\sum_{n \in \mathcal{T}} \frac{r_n^0}{r_n^n} \leq 1,$$

each source $n \in \mathcal{T}$ is scheduled to transmit individually with probability

$$\tilde{f}_n^{\mathrm{u}^\star} = \frac{1}{T}, \quad \forall n \in \mathcal{T},$$

and the probability of concurrent operation is zero, i.e., $\tilde{f}_0^{\mathrm{u}^\star} = 0$.

(2) If

$$\sum_{n \in \mathcal{T}} \frac{r_n^0}{r_n^n} > 1,$$

the optimal policy is of a threshold type with threshold $R(\mathcal{J})$ given by

$$R(\mathcal{J}) = \frac{1 - \sum_{j \in \mathcal{J}} r_j^0 / r_j^j}{T - |\mathcal{J}|}. \tag{4.19}$$

Specifically,

(a) A source $j \in \mathcal{T}$ is scheduled to transmit individually with probability $\tilde{f}_j^{\mathrm{u}^\star} > 0$ (i.e., j is individually activated and belongs in $\mathcal{J}$) given by

$$\tilde{f}_j^{\mathrm{u}\star} = \frac{1}{T}\left(1 - \sum_{i\in\mathcal{J}^c} \frac{r_j^0/r_j^j}{1-\sum_{j\in\mathcal{J}} r_j^0/r_j^j}\right), \tag{4.20}$$

if and only if

$$\frac{r_j^0}{r_j^j} < R(\mathcal{J}). \tag{4.21}$$

(b) All sources operate concurrently with probability $\tilde{f}_0^{\mathrm{u}\star}$ given by

$$\tilde{f}_0^{\mathrm{u}\star} = \frac{T - |\mathcal{J}|}{T\left(1-\sum_{j\in\mathcal{J}} r_j^0/r_j^j\right)}. \tag{4.22}$$

Corollary 4.7. The cardinality of the set $\mathcal{J}$ under the optimal policy specified in Corollary 4.6 is given by the following:

$$|\mathcal{J}| = \underset{\ell\in\{0,1,\dots,T\}}{\arg\max} \frac{1-\sum_{j=1}^{\ell} \tilde{r}_j^0/\tilde{r}_j^j}{T-\ell}. \tag{4.23}$$

4.3.2 Numerical Results

In this section, we analyze the performance of the proposed policies through a set of numerical experiments. First, we consider the special case of purely unicast traffic. Then, we proceed to a more general case that involves both unicast and multicast sessions. Throughout this section, we focus only on the criterion of proportional fairness. To illustrate our results we assume that the data rate $r(\cdot)$ is given by the single user Shannon formula, i.e., $r(\gamma) = \log_2(1+\gamma)$, under the assumption of unit bandwidth. As in Section 3.3 and in [48], we could just as well use the formulas that relate the data rate to the SINR threshold under arbitrary modulation schemes but with additional numerical complexity. Finally, we assume that the duration of a time slot is equal to 1 s.

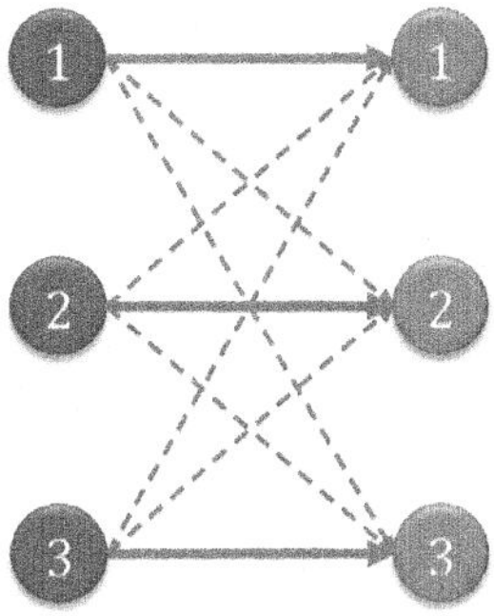

Fig. 4.3 A static network of three source/destination pairs under unicast traffic.

4.3.2.1 Unicast Case

The first wireless network we consider is shown in Figure 4.3. It is a single-hop, static network of three source/destination pairs of unicast data traffic. The maximum transmission powers at the sources are $P_1^{\max} = P/2, P_2^{\max} = P, P_3^{\max} = 6 \times P$, where $P = 6.0 \times 10^{-5}$ W. Furthermore, the power of the thermal noise at each destination is equal to $N_0 = 3.34 \times 10^{-6}$ W. We also assume that the path losses between the three sources and the three destinations are given by (3.10). Under this channel model, we compare the performance of two proportionally fair policies. By proportionally fair we mean that the corresponding probabilities with which the different actions in the set $\mathcal{A}$ are chosen to solve the unicast problem (4.7)–(4.9), obtained by replacing $|\mathcal{D}(n)| = 1$ for each n in (4.7). The first policy we consider is a proportionally fair TDMA scheme that activates a single source at any given time (at its highest possible rate) with a probability optimized to ensure proportional fairness in the effective received rates when only TDMA actions are considered. The second policy is a rate control policy that can choose any among the $2^3 - 1$ possible rate control and scheduling decisions, i.e., activate a single source at any given time, or pairs of sources, or all of them together. Again, the probability with which each action is selected is optimized so that the effective rate at each destination is proportionally fair under this set of actions.

Figure 4.4 shows the variation of the effective proportionally fair rates of the three source/destination pairs under the considered policies

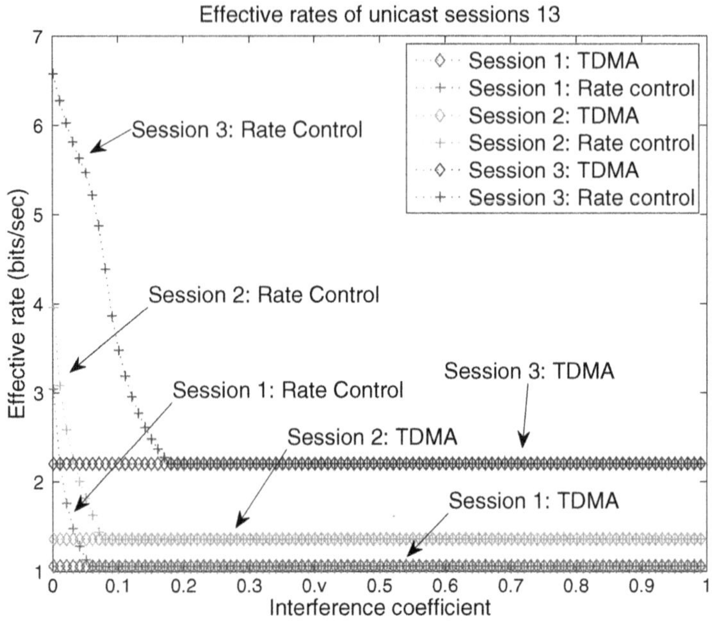

Fig. 4.4 Effective rate of the three unicast sessions with respect to the interference coefficient β.

as the interference level in the system increases. First, we observe that the proportionally fair rate of the third source/destination pair is higher than the corresponding rates of the other two pairs and that the first source/destination pair has the lowest rate under all values of the interference coefficient. This is a natural outcome stemming from the specific selections on the maximum transmission powers of the respective sources. Our second observation confirms our intuitive explanation that the rate control policy performs strictly better than the pure TDMA scheme at low levels of interference, i.e., when the interference coefficient β is small. However, the performance gains of the proportionally fair rate control policy over the proportionally fair TDMA policy diminish rather quickly as β increases. For any interference coefficient $\beta > 0.2$, we observe that the rate control scheme converges to a TDMA scheme and thus both policies achieve the same performance. In other words, above a certain level of interference, a proportionally fair TDMA scheme becomes the optimum choice.

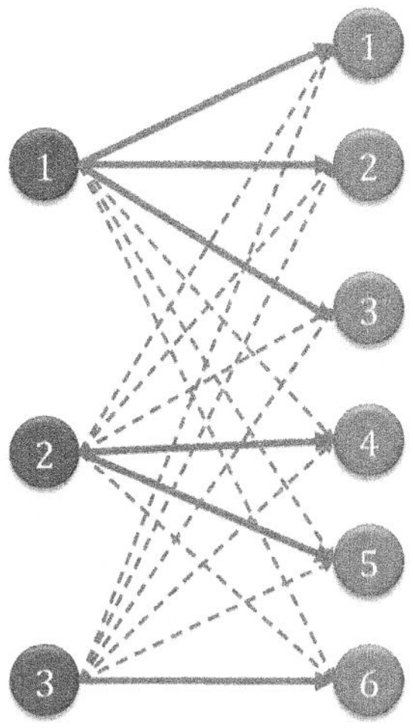

Fig. 4.5 A single-hop network of three sources and six destinations.

4.3.2.2 Multicast Case

In this section, we consider a static, single-hop network with three sources and six destinations as shown in Figure 4.5. The sets of the destinations for each source are $\mathcal{D}(1) = \{1,2,3\}$, $\mathcal{D}(2) = \{4,5\}$, and $\mathcal{D}(3) = \{6\}$; in other words sources 1 and 2 multicast to their respective destination groups while source 3 is a unicast source. We set the maximum transmission powers to be equal, i.e., $P_n^{\max} = P$, $n = 1,2,3$, where $P = 6.0 \times 10^{-5}$ W. As in the previous section, the thermal noise power at each destination is $N_0 = 3.34 \times 10^{-6}$ W.

The path losses between the three sources and the six destinations are captured by the path loss matrix $\mathbf{G}$, given as

$$\mathbf{G} = \begin{bmatrix} 0.8 & 0.9 & 0.75 & \beta & \beta & \beta \\ \beta & \beta & \beta & 0.85 & 0.9 & \beta \\ \beta & \beta & \beta & \beta & \beta & 0.7 \end{bmatrix}.$$

As before, $\beta \in [0,1]$ represents the *interference coefficient.* In Figures 4.6, 4.7, and 4.8, the proportionally fair rates of each multicast session are plotted as a function of the interference coefficient β for three policies. Specifically, we consider (i) a proportionally fair rate control scheme that allows all $2^3 - 1$ possible rate control and scheduling

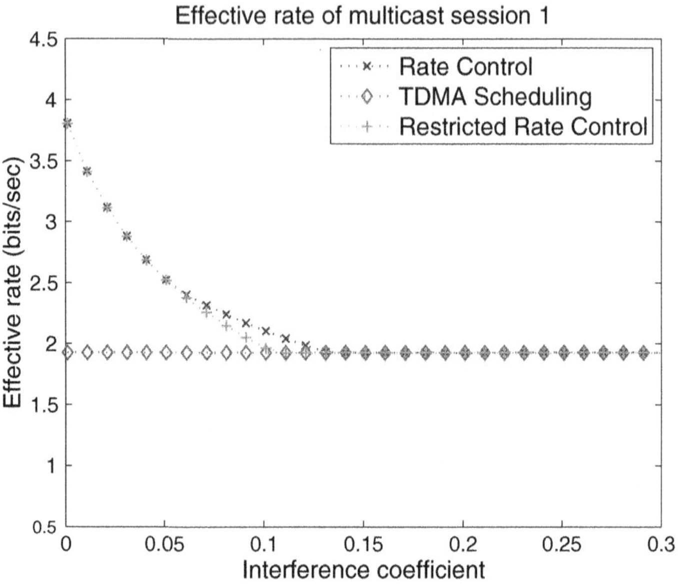

Fig. 4.6 Effective rate of source 1 with respect to the interference coefficient β.

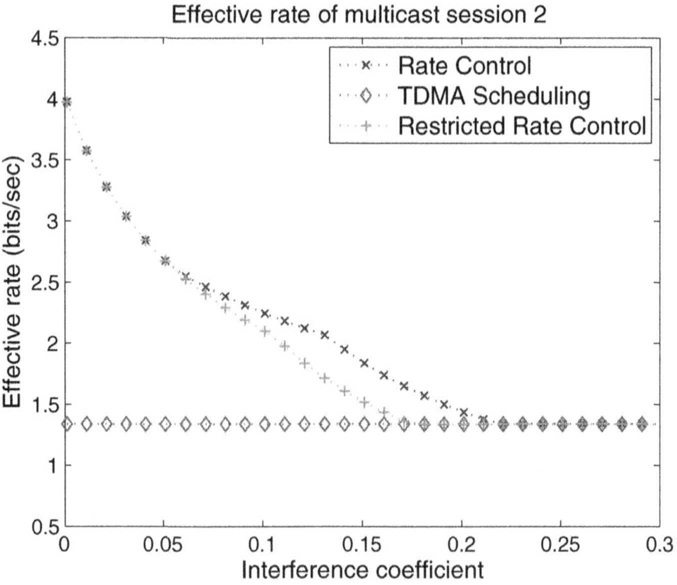

Fig. 4.7 Effective rate of source 2 with respect to the interference coefficient β.

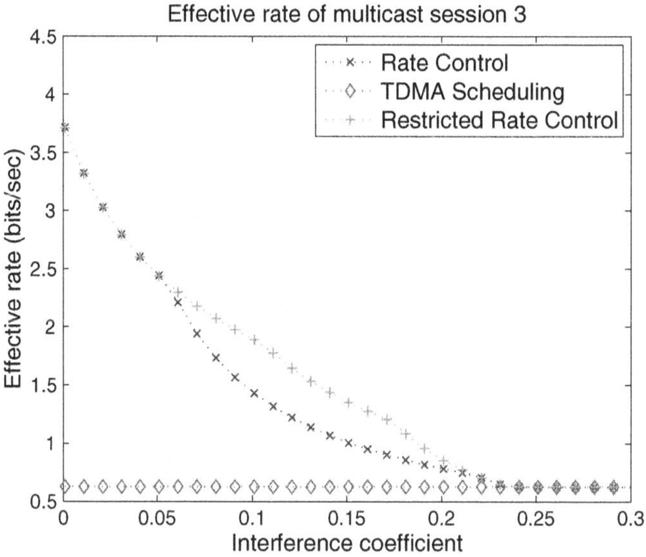

Fig. 4.8 Effective rate of source 3 with respect to the interference coefficient β.

ways of activating the three sources, (ii) a proportionally fair TDMA scheme, where a single source is activated at any given time, and (iii) the restricted rate control scheme that considers either operation "all-at-once" or "one-at-a-time". Note that in our plots we only show the effective rates for values of β between 0 and 0.3 since for larger values of β all three policies converge to the TDMA scheme. As in the unicast experiment, the corresponding action probabilities are optimized so that the effective received rates are proportionally fair.

The maximum rates under the rate control and the restricted rate control schemes correspond to the case where there is no interference between all the multicast sources (i.e., when $\beta = 0$). Under this case, it is easy to see that the optimal action is to allow all sources to transmit simultaneously, i.e., to use operation "all-at-once". This is the reason why the maximum rates for the restricted rate control scheme and for the rate control scheme coincide. Furthermore, one may observe that the rates of each multicast source are not exactly scaled with respect to the destination set sizes. Although the transmission powers for all multicast sources are the same in this multicast scenario, the path loss coefficients between the multicast sources and their corresponding destinations are different for each source as can be seen from the matrix **G**. Note that in this volume the multicast rate of a source is defined to be the minimum rate at which all the destinations in its multicast session can successfully receive the transmission. Given this definition and using the fact that the optimal operation at $\beta = 0$ is operation "all-at-once", the rates under the $\beta = 0$ case (i.e., no interference case) will be proportional to $P \times 0.75$ for multicast source 1, $P \times 0.9$ for multicast source 2, and $P \times 0.7$ for source 3, as can be observed from the matrix **G**. However, as the interference coefficient increases we indeed observe that the rates scale according to the sizes of the destination sets (e.g., for $\beta > 0.25$, the rate ratios are proportional to the size of the destination sets). Similarly to the unicast case, when the levels of interference are low (i.e., β is close to 0), the two proportionally fair rate-control schemes achieve much higher rates than the corresponding TDMA scheme. Furthermore, both rate-control schemes converge fast, as expected, to the TDMA scheduling policy as the interference coefficient β increases. Finally, the fact that the third source has a higher rate under the restricted set of actions compared to the rate-control scheme is not contradictory, and can be explained as follows. Since the optimization criterion is the weighted sum of logarithms of rates over *all* sources, it is possible that under a suboptimal scheme one of the sources receives a higher rate. However, the overall rate received under the rate control scheme will always be higher than the corresponding rate under the restricted one.

4.4 Minimum-Length Scheduling

In this section, we revisit the minimum-length scheduling problem of Section 2. The discrete nature of this problem also makes it combinatorially complex. Thus here, we are going to map the problem given by (2.10)–(2.12) to a continuous time one. Therefore, instead of seeking the minimum number of time slots required to deliver all data traffic to its respective destinations, we are interested to obtain the minimum "duration" or "period of time" that has to elapse until all network queues empty. In this way, the minimum length scheduling problem becomes a linear program with a relatively small number of constraints and a large number of variables as in the formulations of [12], [33], and [34]. In order to solve this linear program, we follow a different approach than [12], [33], and [34]. In particular, we reduce the number of variables involved, i.e., the scheduling and rate control decisions that the policy employs, and then obtain an optimal solution for this reduced problem.

Specifically, we restrict the set of possible actions $\mathcal{A}$ to contain only feasible rate vectors obtained by the two simple schemes described in Section 4.1.1. By doing so, we decrease the cardinality of $\mathcal{A}$ to $T + 1$.

Let us denote by τ_j for $j \in \{0,\ldots,T\}$ the period of time that Action j is utilized. Then, the continuous time equivalent of (2.10)–(2.12) under the reduced space of actions is:

$$\textbf{minimize}: \sum_{j=0}^{T} \tau_j \tag{4.24}$$

$$\textbf{subject to}: \ b_n \leq \tau_n r_n^n + \tau_0 r_n^0, \quad \forall n \in \mathcal{T} \tag{4.25}$$

$$\tau_j \geq 0, \quad \forall j \in \{0,\ldots,T\}. \tag{4.26}$$

The following theorem characterizes an optimal scheduling and rate control policy that solves (4.24)–(4.26).

Proposition 4.8. A minimum-length scheduling and rate control policy solving (4.24)–(4.26) takes actions according to the following:

(1) If

$$\sum_{n=1}^{T} \frac{r_n^0}{r_n^n} \leq 1,$$

Action n is chosen ($n = 1,\ldots,T$) for a duration of

$$\tau_n = \frac{b_n}{r_n^n},$$

while Action 0 is never employed, i.e.,

$$\tau_0 = 0.$$

(2) If

$$\sum_{n=1}^{T} \frac{r_n^0}{r_n^n} \geq 1,$$

then a subset of sources $\mathcal{J}$ is chosen such that for every $n \in \mathcal{J}$, Action n is chosen for a duration of

$$\tau_n = \frac{b_n - \tau_0 r_n^0}{r_n^n},$$

and Action 0 is selected for a period of

$$\tau_0 = \max_{i \in \mathcal{T} \setminus \mathcal{J}} \frac{b_i}{r_i^0}.$$

To completely characterize the policy we need to specify the set $\mathcal{J}$, which results from the following proposition.

Proposition 4.9. Consider an ordering of the sources in decreasing order of the values b_n/r_n^0 for every $n \in \mathcal{T}$. Let the corresponding indexing of the sources be $\{\ell_n\}_{n=1}^{T}$, that is $b_{\ell_1}/r_{\ell_1}^0 \geq \cdots \geq b_{\ell_T}/r_{\ell_T}^0$. Then, the set $\mathcal{J}$ contains those sources with the highest b_n/r_n^0 ratios and the cardinality $|\mathcal{J}|$ of the set $\mathcal{J}$ is given by

$$\begin{aligned} |\mathcal{J}| = & \operatorname*{arg\,min}_{n \in \{0,\ldots,T\}} \left\{ \left(\max_{i \in \mathcal{T} \setminus \{0,\ldots,n\}} \frac{b_{\ell_i}}{r_{\ell_i}^0} \right) \right. \\ & \left. + \sum_{\{j: j \geq 1, j \leq n\}} \frac{b_{\ell_j}}{r_{\ell_j}^{\ell_j}} - \left(\max_{i \in \mathcal{T} \setminus \{0,\ldots,n\}} \frac{b_{\ell_i}}{r_{\ell_i}^0} \right) \sum_{\{j: j \geq 1, j \leq n\}} \frac{r_{\ell_j}^0}{r_{\ell_j}^{\ell_j}} \right\}. \end{aligned} \tag{4.27}$$

The proofs of Proposition 4.8 and 4.9 appear in [47]. From the above we conclude that the set $\mathcal{J}$ contains the sources with the highest $|\mathcal{J}|$ values of b_n/r_n^0, where $|\mathcal{J}|$ is given by Proposition 4.9. Hence, an optimal scheduling and rate control policy individually activates the sources that either have a very high initial demand or whose rates under concurrent operation are very low, e.g., due to excessive amounts of interference caused by other concurrent transmissions. Those sources must be further assisted towards emptying their queues by being granted individual access to the channel.

5

Maximum Stable Throughput

An important criterion to measure performance in a system where "bursty" exogenous arrivals enter the network queues is the rate at which data are delivered to their receivers, while guaranteeing that the queues at the network nodes do not grow without bound. This is what we generally call *stable throughput* of the network. Under stability the throughput rates coincide with the arrival rates of the traffic entering the network. The main objective in this section is to identify the maximum set of achievable stable arrival rates, or otherwise to characterize the *stability region* of a network under imperfect channel information and further provide joint scheduling and routing network control policies that achieve these rates.

There exists a rich literature on the subject of maximum stable throughput (see e.g., [45, 51, 65, 66]). Specifically in [65], a scheduling policy that achieves maximum stable throughput in single-hop time-varying networks is identified. Moreover, in [66], the authors characterize the stability region of static, multi-hop radio networks with multiple commodities, and propose a centralized, stationary, scheduling and routing rule, commonly referred as the "back-pressure", that achieves maximum stable throughput. The back-pressure policy forwards the traffic through the network from queues with high loads to queues with lower loads and achieves stability by load-balancing the queues in the network. Furthermore, the authors in [66] show that their proposed policy is at least as good as any stationary policy. Under the assumption that a scheduled transmission is always successful, they prove that their policy performs at least as well as any non-stationary policy with respect to maximizing the stable throughput region of the network. In fact, the back-pressure algorithm of [66] has been shown to maximize the stable throughput region under a variety of conditions. In previous work of ours, [51, 52], we proved optimality of a policy inspired by the back-pressure algorithm of [66] within the set of all stationary policies in the more general setting of wireless networks with *time-varying topologies.* Furthermore, [51, 52] also differ from [66] in that our proposed policy gives priority to each commodity according to a preassigned commodity weight. In [51, 52], and [66], it is assumed that links are imperfect and that a scheduled transmission may fail, based on a link failure probability, which is independent of the identity, and the number of the simultaneously activated links. Finally, in another related study, [45], a joint scheduling, routing, and power control policy, also inspired by the back-pressure algorithm, is proposed that maximizes the stable throughput region of time-varying wireless networks. The authors in [45] consider a time-varying process of perfect channels, i.e., a transmission through a link is always successful.

However, in practice the channel conditions can only be estimated, and hence exact knowledge of the current channel state is likely to be unavailable. Furthermore, in cases when the channel process varies rapidly with time (e.g., fast fading) or when the propagation delay of the feedback channel is large (see e.g., [39, 75]) this estimate may be highly inaccurate. Hence, the channel state at the time of a scheduling or routing decision can be significantly different from the state at which the actual transmissions take place. The effect of this discrepancy in the channel state may be two-fold; first, certain scheduled transmissions are going to fail, and second, transmissions through certain links that would be successful if scheduled, are not activated. Naturally, this situation will affect the set of stabilizable rates and will result in a smaller stability region that is a subset of the stability region under perfect links or under perfect channel estimation.

Thus in this section, we are going to capture the effect of imperfect channel information and characterize the maximum achievable stable throughput region under uncertainty. We will also obtain a class of policies that maximizes the stable throughput region of *multi-hop, time-varying* networks by having access to only a, perhaps highly inaccurate, *estimate* of the current channel state. This class of policies, which is inspired by the back-pressure idea of [66], is optimal within a broad class of stationary, and non-stationary, even anticipative policies.

5.1 Model Formulation

In this section, we consider multi-hop wireless networks where each network node $n \in \mathcal{M}$ either transmits at a fixed power level $P_n^{\max}$ or remains silent. Furthermore, instead of saturated sources of traffic or sources with a finite traffic load, in this section we focus on bursty traffic. Moreover, we assume that whenever a node transmits it can transmit a fixed number of bits, i.e., here we do not consider rate control and the SINR thresholds are fixed. We assume that a packet consists of as many bits as there are sent with each transmission attempt.

We denote by $\mathcal{H} = \{1,2,\dots,H\}$ the set of distinct commodities of traffic with packet lengths equal to one time slot. The number of exogenous packet arrivals of commodity $h \in \mathcal{H}$ at node n during time slot t is denoted by $A_{nh}(t)$. We let $\mathbf{A}^h(t)$ denote the M-vector $(A_{nh}(t) : n = 1,2,\dots,M)$ of arrivals of the hth commodity during time slot t at every node in the network and $\mathbf{A}(t)$ denote the $M \times H$ matrix $(A_{nh}(t), n = 1,2,\dots,M, h = 1,2,\dots,H)$ of arrivals in time slot t at every node n and for every commodity h. In this section, we also focus on anycast traffic, i.e., traffic that originates from a source and is destined for *any* node within a set of destinations. Traffic of commodity $h \in \mathcal{H}$ is routed in a multi-hop fashion through the network until it reaches any node in a set of *destination nodes* for that commodity, $\Delta_h \subset \mathcal{M}$, where it exits the network. For any commodity $h' \neq h$, the sets $\Delta_{h'}$, and Δ_h may overlap. We further assume that there are no exogenous arrivals of a particular commodity at the destination nodes of that commodity, i.e., $A_{nh}(t) = 0$ for all $n \in \Delta_h$, $h \in \mathcal{H}$.

At each node n there exist H infinite capacity buffers, each holding separately the packets of a particular commodity $h \in \mathcal{H}$ that have reached node n. We denote the queue size for commodity h at node n *at the end* of time slot t by $X_{nh}(t)$. At time slot 0 the queue sizes at all nodes are arbitrary but finite, i.e., $X_{nh}(0) \geq 0$ for every node $n \in \mathcal{M}$, and commodity $h \in \mathcal{H}$. Moreover, the queue size at each destination node $n \in \Delta_h$ of some commodity h and for all time slots $t \geq 0$ satisfies $X_{nh}(t) = 0$. Finally, for every commodity $h \in \mathcal{H}$ we denote by $\mathbf{X}^h(t)$ the M-vector $(X_{nh}(t), n = 1,2,\ldots M)$ of queue sizes of the hth commodity at every node in the network at the end of time slot t and by $\mathbf{X}(t)$ the $M \times H$ matrix $(X_{nh}(t), n = 1,2,\ldots M, h = 1,2,\ldots H)$ of queue sizes of every commodity at every node in the network at the end of time slot t. The set of possible values of $\mathbf{X}(t)$, i.e., the state space of the process $\{\mathbf{X}(t)\}_{t=0}^{\infty}$, is denoted by $\mathcal{X}$.

The channel process $\{\mathbf{G}(t)\}_{t=1}^{\infty}$ defines the channel conditions between every pair of nodes in the network and is assumed to change only at the beginning of each time slot t. Specifically, at time slot t, the channel state $\mathbf{G}(t) = \{(G_{(n,m)}(t), N_0), \forall n,m \in \mathcal{M}\}$ is characterized by the path loss $G_{(n,m)}(t)$ between each pair of nodes n, m, as well as the noise power, N_0. A fundamental aspect of our model that contrasts it from prior work of [51, 52, 65], and [45] is that at the beginning of each time slot t the network controller has access only to an *estimate* $\hat{\mathbf{G}}(t) = \{(\hat{G}_{(n,m)}(t), \hat{N}_0(t)), \forall n,m \in \mathcal{M}\}$ of the current channel state $\mathbf{G}(t)$. The *estimated* channel state $\hat{\mathbf{G}}(t)$ during slot t is characterized by the *estimated* path loss $\hat{G}_{(n,m)}(t)$ between each pair of nodes n, m, and the *estimated* noise power $\hat{N}_0(t)$. Note that although the noise power N_0 is time invariant, its estimate $\hat{N}_0(t)$ depends on time, since as time progresses we may naturally get a monotonically improving estimate.

We further assume that the state space of the *true* and *estimated* channel processes is a finite set of cardinality $|\mathbf{G}|$, which is naturally assumed to be common for both $\{\mathbf{G}(t)\}_{t=1}^{\infty}$, and $\{\hat{\mathbf{G}}(t)\}_{t=1}^{\infty}$. For example, that would be the case if we consider node mobility that is restricted to occur only among points of a finite grid. We denote this common set by $\mathcal{G} = \{\mathbf{G}^{(1)}, \mathbf{G}^{(2)}, \ldots, \mathbf{G}^{(|\mathbf{G}|)}\}$. We will further denote by $\mathcal{I} = \{1, 2, \ldots, |\mathbf{G}|\}$ the set of indices that label the elements of $\mathcal{G}$.

At every time slot t, a (unidirectional) link $\ell = (n,m)$ from node n to node m under the true channel state $\mathbf{G}(t) \in \mathcal{G}$ is defined to exist, if the Signal to Noise Ratio (SNR) at m exceeds a certain, non-negative, threshold γ_m, i.e.,

$$\mathrm{SNR}(\ell,t) = \frac{P_n^{\max} \, G_{(n,m)}(t)}{N_0} \geq \gamma_m. \tag{5.1}$$

We denote the source node n of link ℓ by $s(\ell)$ and its receiver node m by $d(\ell)$. Given the time variability of the channel conditions and the fact that nodes are mobile, the total number of links, L, can be as large as $M \times (M-1)$. We denote by $\mathcal{L} = \{1,2,\ldots,L\}$ the set of indices of all links in the network.

The fact that the wireless medium is a shared resource poses limitations on the set of nodes that may successfully transmit simultaneously. Hence, not every subset of links in $\mathcal{L}$ can be concurrently activated. In order to take the physical layer access constraints into account, appropriate medium access control schemes need to be introduced. Towards this end, we will say that a vector $\mathbf{c}$ is a valid activation vector with respect to some channel state $\mathbf{G}(t)$ if it is an L-element binary vector, each entry of which corresponds to a (unidirectional) link. At every time slot t, the entries of this vector are equal to one for those links that are concurrently activated at time slot t and zero for all other links. A valid activation vector is such that for every link $\ell \in \mathcal{L}$ such that the ℓth entry c_ℓ of $\mathbf{c}$ satisfies $c_\ell = 1$, the SINR criterion

$$\mathrm{SINR}^{\,\mathbf{c}}(\ell,t) = \frac{P_{s(\ell)}^{\max} \, G_{(s(\ell),d(\ell))}(t)}{N_0 + \sum_{\substack{\ell' \in \mathcal{L} \setminus \{\ell\} \\ \text{s.t. } c_{\ell'} = 1}} P_{s(\ell')}^{\max} \, G_{(s(\ell'),d(\ell))}(t)} \geq \gamma_{d(\ell)}, \tag{5.2}$$

is satisfied with $c_{\ell'}$ being the ℓ'th entry of $\mathbf{c}$. Note that this definition is different from the definition of a valid activation vector in Section 3.1, where an activation vector was assumed to determine the active sources (rather than the active links). With a careful observation of the two definitions it is easy to see that they yield an equivalent scheduling result. Note also that here we abuse notation slightly as the SINR criterion is parameterized by the binary vector $\mathbf{c}$ while in Section 1 the corresponding expression (1.1) was parameterized by the transmission powers of the sources in the network. Again, it is easy to see that both SINR expressions are equivalent. The criterion of (5.2) implies that the corresponding transmissions through all links $\ell \in \mathcal{L}$ with $c_\ell = 1$ will be successful under channel state $\mathbf{G}(t)$. Similarly, the estimated SINR criterion under $\hat{\mathbf{G}}(t)$ can be written as

$$\widehat{\text{SINR}}^{\,\mathbf{c}}(\ell,t) = \frac{P^{\max}_{s(\ell)}\ \hat{G}_{(s(\ell),d(\ell))}(t)}{\hat{N}_0(t) + \sum_{\substack{\ell' \in \mathcal{L} \setminus \{\ell\} \\ \text{s.t. } c_{\ell'} = 1}} P^{\max}_{s(\ell')}\ \hat{G}_{(s(\ell'),d(\ell))}(t)} \geq \gamma_{d(\ell)}. \qquad (5.3)$$

Note that due to the inaccuracy of the estimate, an activation vector selected at time slot t may be valid with respect to the estimated channel state $\hat{\mathbf{G}}(t)$ at slot t but not valid with respect to the true channel state $\mathbf{G}(t)$, and vice versa.

For every possible channel state $\mathbf{G}^{(k)} \in \mathcal{G}$ where $k \in \mathcal{I}$, we denote by Γ_k the *constraint set* of $\mathbf{G}^{(k)}$, i.e., the set of all *valid* activation vectors with respect to $\mathbf{G}^{(k)}$. Note that for every activation vector $\mathbf{c}' \in \{0,1\}^L$ that is componentwise smaller than some vector $\mathbf{c} \in \Gamma_k$, i.e., $\mathbf{c}' \leq \mathbf{c}$, it follows that $\mathbf{c}' \in \Gamma_k$. This is natural because for every collection of links that jointly satisfy the SINR criteria of (5.2)–(5.3), these criteria will still be satisfied by switching off certain transmissions. From the above observation it follows trivially that for every $k \in \mathcal{I}$ the $\mathbf{0}$-vector is also a valid activation vector for each channel state $\mathbf{G}^{(k)} \in \mathcal{G}$.

For each commodity h, consider a process $\{\mathbf{E}^h(t)\}_{t=1}^{\infty}$ that for every time slot t gives the link activations for packets of commodity h. In other words for every time slot t the vector $\mathbf{E}^h(t)$ is an L-element binary vector, the entries of which are equal to one for those links that are simultaneously activated and packets of commodity h are transmitted through them and are equal to zero otherwise. Furthermore, for every time slot t we define $\mathbf{E}(t) = \sum_{h=1}^{H} \mathbf{E}^h(t)$. The process $\{\mathbf{E}(t)\}_{t=1}^{\infty}$ corresponds to the overall link activations for every time slot t and it is such that whenever at time slot t the estimated channel process is in state $\mathbf{G}^{(k)}$, the vector $\mathbf{E}(t)$ is a valid link activation vector with respect to $\mathbf{G}^{(k)}$. This means that $\mathbf{E}(t)$ is a vector from the constraint set Γ_k, i.e, $\mathbf{E}(t) \in \Gamma_k$. We call the process $\{\mathbf{E}^h(t)\}_{t=1}^{\infty}$ an *activation process.* Recall that the constraint set has the property that for every vector in the constraint set, any other vector that is smaller component-wise must be in the constraint set as well. Since $\mathbf{E}(t) \in \Gamma_k$, the aforementioned property implies that for every commodity h the corresponding vector $\mathbf{E}^h(t)$ is also a valid activation vector with respect to $\mathbf{G}^{(k)}$, i.e., it satisfies $\mathbf{E}^h(t) \in \Gamma_k$. Furthermore, we require that for each commodity h, a vector $\mathbf{E}^h(t)$ must be such that its ℓth component, $(\mathbf{E}^h(t))_\ell$, takes the value zero for all those time slots t that the queue size at source node of the link, $s(\ell)$, for commodity h is equal to zero at the time of the link activation, i.e., $X_{s(\ell)h}(t-1) = 0$. We say that every such process $\{\mathbf{E}(t)\}_{t=1}^{\infty}$ is an *admissible policy*, and the process $\{\mathbf{E}^h(t),\ h \in \mathcal{H}\}_{t=1}^{\infty}$ is an *admissible policy corresponding to the hth commodity.* Unless otherwise specified all the policies we consider are valid.

Furthermore, for every time slot t where $\hat{\mathbf{G}}(t) = \mathbf{G}^{(k)}$ for some $k \in \mathcal{I}$ and for every activation vector $\mathbf{c} \in \Gamma_k$, we construct the $L \times L$ diagonal indicator matrix $\mathbf{Q}^{\mathbf{c}}(t)$, whose ℓth diagonal entry, $(\mathbf{Q}^{\mathbf{c}}(t))_\ell$, satisfies

$$(\mathbf{Q}^{\mathbf{c}}(t))_\ell = \begin{cases} 1, & \text{if}\big(\mathrm{SINR}^{\mathbf{c}}(\ell,t) \geq \gamma_{d(\ell)},\ \widehat{\mathrm{SINR}^{\mathbf{c}}}(\ell,t) \geq \gamma_{d(\ell)}\big) \text{ or} \\ & \quad \big(\mathrm{SINR}^{\mathbf{c}}(\ell,t) < \gamma_{d(\ell)},\ \widehat{\mathrm{SINR}^{\mathbf{c}}}(\ell,t) < \gamma_{d(\ell)}\big), \\ 0, & \text{otherwise.} \end{cases} \tag{5.4}$$

Intuitively, for any given activation vector $\mathbf{c} \in \Gamma_k$ and estimated channel state $\mathbf{G}^{(k)}$, the ℓth entry of the matrix $\mathbf{Q}^{\mathbf{c}}(t)$ takes the value 1 only when the estimator estimates the channel correctly in the sense that the values of the corresponding SINRs under both the *true* and *estimated* channel state lie on the same side of the inequality. Note that whether $(\mathbf{Q}^{\mathbf{c}}(t))_\ell$ is equal to 1 or 0 depends on the overall link activations given by the vector $\mathbf{c}$. In the ideal case of perfect channel estimation, the matrix $\mathbf{Q}^{\mathbf{c}}(t)$ is the identity matrix, i.e., $\mathbf{Q}^{\mathbf{c}}(t) = \mathrm{I}$, for every time slot t where the estimated channel state is in state $\mathbf{G}^{(k)}$ for some $k \in \mathcal{I}$ and for every activation vector $\mathbf{c} \in \Gamma_k$.

Also, for every commodity h we define the matrix $\mathbf{R}^h$ as an $M \times L$ matrix that denotes the changes in the queue sizes after a successful link activation. The (n,ℓ) entry, $R^h_{n\ell}$, of this matrix equals

$$R^h_{n\ell} = \begin{cases} 1, & \text{if } n = d(\ell) \notin \Delta_h, \\ -1, & \text{if } n = s(\ell), \\ 0, & \text{otherwise.} \end{cases} \tag{5.5}$$

Note that $R^h_{n\ell} = 0$ when $n = d(\ell) \in \Delta_h$, as packets of commodity h arriving at n exit the system. Overall, the above yields the following dynamic equation for the queue sizes

$$\mathbf{X}^h(t+1) = \mathbf{X}^h(t) + \mathbf{R}^h\, \mathbf{Q}^{\mathbf{E}(t+1)}(t+1)\, \mathbf{E}^h(t+1) + \mathbf{A}^h(t+1), \;\; t \geq 0. \tag{5.6}$$

Throughout this volume we make use of the following assumption on the input processes:

Assumption 2. (a) The triplet $\{\mathbf{G}(t), \hat{\mathbf{G}}(t), \mathbf{A}(t)\}_{t=1}^{\infty}$ is i.i.d. over time and independent of $\mathbf{X}(0)$. (b) The arrival process has finite second moments, i.e., $\mathbb{E}[\mathbf{A}(t)^2] < \infty$.

Assumption 2(a) guarantees that each of the processes $\{\mathbf{G}(t)\}_{t=1}^{\infty}$, $\{\hat{\mathbf{G}}(t)\}_{t=1}^{\infty}$, and $\{\mathbf{A}(t)\}_{t=1}^{\infty}$ are individually i.i.d. and hence have a stationary distribution. In particular, the probability $p_{\hat{\mathbf{G}}}(k)$ of the occurrence of *estimated* channel state $\mathbf{G}^{(k)} \in \mathcal{G}$, given by

$$p_{\hat{\mathbf{G}}}(k) = \text{Prob}[\hat{\mathbf{G}}(t) = \mathbf{G}^{(k)}], \quad \forall k \in \mathcal{I}, \tag{5.7}$$

does not depend on t. Without loss of generality, we assume that

$$p_{\hat{\mathbf{G}}}(k) > 0, \quad \forall\, k \in \mathcal{I}. \tag{5.8}$$

Indeed, all our results are probabilistic in nature and are not affected if we discard sample paths corresponding to a nullset of outcomes. Moreover, from Assumption 2(a) it follows that although the processes are i.i.d. in time, for any particular time slot t they can be correlated among themselves. For example, the true and estimated channel states, $\mathbf{G}(t)$ and $\hat{\mathbf{G}}(t)$, are naturally correlated but not $\mathbf{G}(t)$ and $\hat{\mathbf{G}}(t-1)$.

From Assumption 2(b), it follows that the first moments of the arrival process $\{\mathbf{A}(t)\}_{t=1}^{\infty}$ are also finite, i.e., $\lambda_{nh} = \mathbb{E}[A_{nh}(t)]$, where the quantity λ_{nh} corresponds to the arrival rate of commodity h at node n. We also denote by $\boldsymbol{\lambda}$ the *arrival rate matrix* (λ_{nh}, $n = 1,2,\ldots,M$, $h = 1,2,\ldots,H$) of arrival rates at every node in the network, and for every commodity. Finally, for each commodity $h \in \mathcal{H}$ we write $\boldsymbol{\lambda}^h$ for the M-vector $\boldsymbol{\lambda}^h = (\lambda_{nh},\, n = 1,2,\ldots,M)$ of arrivals of the hth commodity at

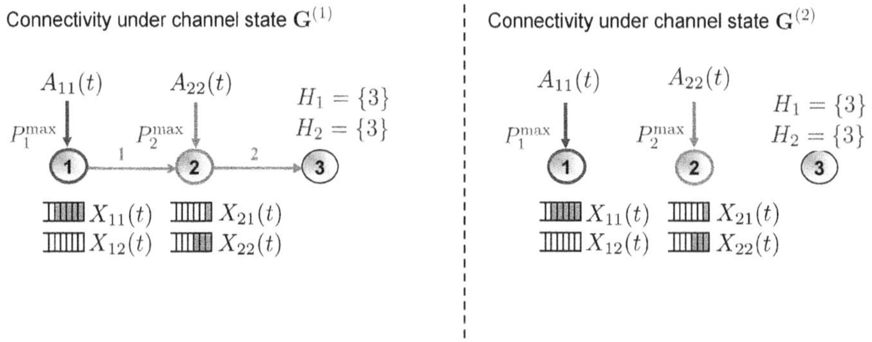

Fig. 5.1 The possible connectivities of a three-node network under two possible channel states, $\mathbf{G}^{(1)}$ and $\mathbf{G}^{(2)}$.

every node in the network. All arrival rates in our model are measured in terms of packets per time slot.

The nomenclature defined so far is summarized through an example in Figure 5.1, where we consider a network of three nodes, i.e., $\mathcal{M} = \{1,2,3\}$. Nodes 1 and 2 transmit at a fixed powers $P_1^{\max}$ and $P_2^{\max}$ respectively. We consider that the channel conditions are such that we have two possible channel states, namely $\mathcal{G} = \{\mathbf{G}^{(1)}, \mathbf{G}^{(2)}\}$. On the left side of the figure, we give the possible links that can be established under channel state $\mathbf{G}^{(1)}$ and on the right side of the figure we give the set of possible links under channel state $\mathbf{G}^{(2)}$. Specifically, when the channel state is $\mathbf{G}^{(1)}$, there exist two possible links, namely links 1 and 2, where a "link" satisfies the SNR criterion of (5.1) and when it is $\mathbf{G}^{(2)}$ no connectivity exists among the nodes. Hence, $\mathcal{L} = \{1,2\}$. Furthermore, although both links 1 and 2 are in $\mathcal{L}$, we assume that they cannot be activated simultaneously due to the fact that they do not jointly satisfy the physical layer constraints of SINR. Specifically, we assume that at most one of them can be activated at any given time. Since the constraint set Γ_k for channel state $\mathbf{G}^{(k)}$ contains all the valid activation vectors with respect to $\mathbf{G}^{(k)}$, we have that $\Gamma_1 = \{[0,0],[0,1],[1,0]\}$ and $\Gamma_2 = \{[0,0]\}$. There exist two commodities of traffic in the network, i.e., $\mathcal{H} = \{1,2\}$. The quantities $A_{11}(t)$ and $A_{22}(t)$ denote the arrivals in packets per slot during time slot t of commodity 1 at node 1, and of commodity 2 at node 2 respectively. We assume that packets of each commodity exit the network at node 3, i.e., $\Delta_h = \{3\}$, for $h = 1,2$. At every node in the network, there exist two infinite capacity buffers that hold separately the packets of each commodity. We indicate the queue size of commodity 1 at node 2 at the end of time slot t by $X_{21}(t)$, and the queue size of commodity 2 at the same node by $X_{22}(t)$. Note that, due to the estimation errors, the policy may schedule, e.g., link 1 assuming that the current channel state is $\mathbf{G}^{(1)}$ when in fact the current state is $\mathbf{G}^{(2)}$ and hence the scheduled transmission through link 1 will fail.

5.2 Stable Throughput Maximization Under Channel State Uncertainty

In this section, we consider a policy that maximizes the stable throughput region of the network by making use of *only* an estimate of the true channel state. Our policy is built upon the back-pressure idea in [66]. As its name suggests, this policy attempts to maximize the stable throughput by spreading the traffic from the more congested to the less congested areas in the network. Accordingly, the policy we introduce activates the nodes of the network in a way that the weighted queue sizes for every commodity h will be kept as close to equal as possible, while at the same time the constraints imposed by the physical layer are being satisfied. Since the physical layer information available to our policy is limited due to the uncertainty in the channel state, our policy will try to maximize the stable throughput region of the network, within a broad class of policies, while having access to only an estimate of the channel conditions.

The routing component of the introduced policy resembles the so-called "hot-potato" routing approach in which nodes simply unload packets to neighboring nodes with smaller queue loads [5]. In fact, in our model, the route any packet follows is determined by the link activation schedule that aims at maximizing the stable throughput region of the network. Hence, although an individual packet may follow a circuitous route towards one of its destination nodes, the overall characteristics of the routes are expected to be reasonable, albeit non-optimal. Since our objective is to achieve maximum stable throughput, this sort of routing is legitimate. No other routing will increase the stable throughput region, although it may decrease the delay that packets of the different commodities experience in the network.

The introduced policy $\boldsymbol{\pi}_0^{\mathbf{w}}$ is parameterized by a weight assignment $\mathbf{w} = (w_h, h = 1, 2, \ldots, H)$, where w_h is a positive weight assigned to each commodity h. Packets corresponding to a commodity of a larger weight are given priority over the others. For every given weight vector $\mathbf{w}$, the *stationary policy* $\mathbf{E}(t) = \boldsymbol{\pi}_0^{\mathbf{w}}(t)$ is a certain H-tuple of mappings $\boldsymbol{\pi}_0^{\mathbf{w}h}\colon \mathcal{X} \times \mathcal{G} \to \{0,1\}^L$, each corresponding to a commodity h and where $\mathbf{E}^h(t) = \boldsymbol{\pi}_0^{\mathbf{w}h}(t)$. So, we also have that $\boldsymbol{\pi}_0^{\mathbf{w}} = \sum_{h=1}^{H} \boldsymbol{\pi}_0^{\mathbf{w}h}$. For every time slot t, the quantity $\boldsymbol{\pi}_0^{\mathbf{w}h}(t)$ indicates the link activations for packets of commodity h and $\boldsymbol{\pi}_0^{\mathbf{w}}(t)$ gives the overall link activations in the network.

We proceed by specifying the stable throughput maximizing policy $\boldsymbol{\pi}_0^{\mathbf{w}}$ in detail. Given the current queue size matrix $\mathbf{x} \in \mathcal{X}$, weight assignment $\mathbf{w}$, and activation vector $\mathbf{c} \in \Gamma_k$ for every estimated channel state $\mathbf{G}^{(k)}$ let

$$\mathbf{D}_{k\mathbf{c}}^{\mathbf{w}h}(\mathbf{x}) = -w_h\, \tilde{\mathbf{Q}}_k^{\mathbf{c}}\, \mathbf{R}^{h\top}\mathbf{x}^h, \quad k \in \mathcal{I}, \quad h \in \mathcal{H}, \quad \mathbf{c} \in \Gamma_k, \tag{5.9}$$

where

$$\tilde{\mathbf{Q}}_k^{\mathbf{c}} = \mathbb{E}\big[\mathbf{Q}^{\mathbf{c}}(t) \mid \hat{\mathbf{G}}(t) = \mathbf{G}^{(k)}\big]. \tag{5.10}$$

From this definition it follows that the matrix $\tilde{\mathbf{Q}}_k^{\mathbf{c}}$ is an $L \times L$ diagonal matrix. Its ℓth diagonal entry, $(\tilde{\mathbf{Q}}_k^{\mathbf{c}})_\ell$, gives the conditional probability that both the estimated and true SINR values corresponding to ℓ lie at the same side of the inequality, provided that the overall link activations in the network are determined through the activation vector $\mathbf{c}$ and the estimated channel state is $\mathbf{G}^{(k)}$. For any given link ℓ, our model allows this probability to be dependent on the concurrent transmissions. For example, this probability is expected to be higher when link ℓ is the only link activated than when link ℓ is activated along with other concurrent nearby transmissions. Also, Assumption 2(a) guarantees that the matrix $\tilde{\mathbf{Q}}_k^{\mathbf{c}}$ for every $k \in \mathcal{I}$ and $\mathbf{c} \in \Gamma_k$, defined in (5.10), is time invariant.

Since the queue size x_{nh} is equal to zero whenever $n \in \Delta_h$, it follows that the ℓth component $\big(\mathbf{D}_{k\mathbf{c}}^{\mathbf{w}h}(\mathbf{x})\big)_\ell$ of $\mathbf{D}_{k\mathbf{c}}^{\mathbf{w}h}(\mathbf{x})$ is the weighted queue size difference

$$\left(\mathbf{D}^{\mathbf{w}h}_{k\mathbf{c}}(\mathbf{x})\right)_{\ell} = w_h(\tilde{\mathbf{Q}}^{\mathbf{c}}_k)_{\ell}(x_{s(\ell)h} - x_{d(\ell)h}). \tag{5.11}$$

For every link $\ell \in \mathcal{L}$, let

$$(\mathbf{D}^{\mathbf{w}}_{k\mathbf{c}}(\mathbf{x}))_{\ell} = \max_{h\in\mathcal{H}} \left(\mathbf{D}^{\mathbf{w}h}_{k\mathbf{c}}(\mathbf{x})\right)_{\ell}, \tag{5.12}$$

and

$$\mathbf{D}^{\mathbf{w}}_{k\mathbf{c}}(\mathbf{x}) = ((\mathbf{D}^{\mathbf{w}}_{k\mathbf{c}}(\mathbf{x}))_{\ell}, \quad \ell = 1,\ldots,L). \tag{5.13}$$

Finally, define

$$(h^{\star}_k(\mathbf{x}))_{\ell} = \arg\max_{h\in\mathcal{H}} \left\{\left(\mathbf{D}^{\mathbf{w}h}_{k\mathbf{c}}(\mathbf{x})\right)_{\ell}\right\}, \tag{5.14}$$

to be the maximizer in (5.12), and also let

$$\mathbf{c}^{\star}_k(\mathbf{x}) = \arg\max_{\mathbf{c}\in\mathcal{T}_k} \left\{\mathbf{D}^{\mathbf{w}}_{k\mathbf{c}}(\mathbf{x})^{\top}\mathbf{c}\right\}. \tag{5.15}$$

Recall that the entries of every valid activation vector $\mathbf{c} \in \Gamma_k$ are either 0 or 1, with 1 indicating activation of the corresponding link. Hence $\mathbf{D}^{\mathbf{w}}_{k\mathbf{c}}(\mathbf{x})^{\top}\mathbf{c}$ is a partial sum of weighted queue size differences over all the links, maximized over all the elements of the constraint set Γ_k. If there exist more than one maximizer in (5.15) ties are resolved arbitrarily provided that a link ℓ will be left inactive whenever the corresponding maximum weighted difference associated with that link is 0. Furthermore, if there exist more than one maximizer in (5.14), ties are resolved arbitrarily. With the above in hand and in the spirit of the optimal policy of [66] the policy $\boldsymbol{\pi}^{\mathbf{w}}_0$ is such that its ℓth entry $(\boldsymbol{\pi}^{\mathbf{w}h}_0(\mathbf{x}, \mathbf{G}^{(k)}))_{\ell}$ is given by

$$(\boldsymbol{\pi}^{\mathbf{w}h}_0(\mathbf{x}, \mathbf{G}^{(k)}))_{\ell} = \begin{cases} 1, & h = (h^{\star}_k(\mathbf{x}))_{\ell}, (\mathbf{c}^{\star}_k(\mathbf{x}))_{\ell} = 1, \quad \text{and} \\ & \quad x_{s(\ell)h} > 0, \\ 0, & \text{otherwise}, \end{cases} \tag{5.16}$$

where $(\mathbf{c}^{\star}_k(\mathbf{x}))_{\ell}$ is the ℓth entry of the vector $\mathbf{c}^{\star}_k(\mathbf{x})$. When a link ℓ is activated, i.e., $\left(\boldsymbol{\pi}^{\mathbf{w}}_0(\mathbf{x}, \mathbf{G}^{(k)})\right)_{\ell} = 1$, the policy $\boldsymbol{\pi}^{\mathbf{w}}_0$ will select for transmission through that link a packet of one of the classes h that achieves the "max" in (5.14).

Note that the matrix $\tilde{\mathbf{Q}}_k^{\mathbf{c}}$ is all the information our policy has regarding the current channel conditions as shown through (5.11), (5.14), and (5.15). The policy employs this information by giving a higher preference to those links for which both the true and the estimated SINRs lie at the same side of the inequality. Specifically, the policy will have the tendency to activate links that have a higher chance of successful transmission.

Clearly, for every commodity h we have that $\boldsymbol{\pi}_0^{\mathbf{w}h}(\mathbf{x},\mathbf{G}^{(k)}) \in \Gamma_k$. Note further that for every link ℓ that is activated, a packet of a single commodity h is transmitted, and hence there will exist a single $\boldsymbol{\pi}_0^{\mathbf{w}h}(\mathbf{x},\mathbf{G}^{(k)})$ that satisfies $(\boldsymbol{\pi}_0^{\mathbf{w}h}(\mathbf{x},\mathbf{G}^{(k)}))_\ell = 1$. From this observation it follows that $\boldsymbol{\pi}_0^{\mathbf{w}}(\mathbf{x},\mathbf{G}^{(k)}) \in \Gamma_k$. The above, along with the fact that the policy leaves a link ℓ inactive whenever the maximum weighted difference over that link is 0, guarantees that $\boldsymbol{\pi}_0^{\mathbf{w}}$ satisfies the conditions for being an admissible policy. In Section 5.4, we will introduce the maximizing property of this policy under the following mild assumption.

Assumption 3. Let $n' \in \mathcal{M}$ be a node such that for some $n \in \mathcal{M}$, $h \in \mathcal{H}$ with $\lambda_{nh} > 0$ there exists a sequence of links $\{\ell_i\}_{i-1}^{\bar{p}} \in \mathcal{L}$, with $s(\ell_1) = n$, $d(\ell_i) = s(\ell_{i+1})$, $i = 1,\dots,\bar{p}-1$, and $d(\ell_{\bar{p}}) = n'$ such that $\forall\, i = 1,\dots,\bar{p}$

$$\text{Prob}[\text{SNR}(\ell_i,t) \geq \gamma_{d(\ell_i)} \quad \text{and} \quad \widehat{\text{SNR}}(\ell_i,t) \geq \gamma_{d(\ell_i)}] > 0, \tag{5.17}$$

where $\text{SNR}(\ell,t)$ is obtained by (5.1) and $\widehat{\text{SNR}}(\ell,t)$ is defined similarly as

$$\widehat{\text{SNR}}(\ell,t) = \frac{P_{s(\ell)}^{\max}\, \hat{G}_{(s(\ell),d(\ell))}(t)}{\hat{N}_0(t)}. \tag{5.18}$$

Then, there exists a node $n'' \in \Delta_h$, and a sequence of links $\{\ell_i'\}_{i=1}^{\bar{p}'} \in \mathcal{L}$ with $s(\ell_1') = n'$, $d(\ell_i') = s(\ell_{i+1}')$, $i = 1,\dots,\bar{p}'-1$, and $d(\ell_{\bar{p}'}') \in \Delta_h$ such that (5.17) holds with $\{\ell_i\}_{i=1}^{\bar{p}}$ replaced by $\{\ell_i'\}_{i=1}^{\bar{p}'}$.

Assumption 3 is an assumption on sufficient connectivity of the network. Specifically it requires that for any node that may receive traffic of a particular commodity, there should also exist a downstream path of links to some destination node for that commodity under both the true and estimated channel states.

5.2.1 System Stability

The state of our system is driven by the process of the queue sizes. In this section, we show that under Assumption 2(a) and policy $\pi_0^{\mathbf{w}}$, the queue size process defined by (5.6), i.e., the state of our system, evolves according to a homogeneous Markov Chain. Our aim is to show that this Markov Chain is stable and thus derive network stability for as large a set of arrival rates as possible.

Proposition 5.1. Under Assumption 2(a), the process $\{\mathbf{X}(t)\}_{t=0}^{\infty}$ generated by (5.6) with $\mathbf{E}^h(t) = \pi_0^{\mathbf{w}h}(\mathbf{X}(t-1), \hat{\mathbf{G}}(t))$ for every $h \in \mathcal{H}$ is a homogeneous Markov Chain. Furthermore, $\mathbf{X}(t)$ is independent of $(\mathbf{G}(t'), \hat{\mathbf{G}}(t'), \mathbf{A}(t'))$ for all $t' > t \geq 0$.

The result in the above proposition is a direct consequence of the fact that any process defined by a recurrence equation driven by white noise input, with initial value independent of the input, is Markov (see, e.g., [14, Theorem 2.1]).

A usual definition for stability of an irreducible Markov Chain is that the Markov Chain is positive recurrent. When the Markov Chain is not guaranteed to be irreducible, a more general definition for stability needs to be employed. Following [66], we adopt the following definition for stability of a (not necessarily irreducible) homogeneous Markov Chain.

Definition 5.1. [66] Let $\{Y(t)\}_{t=0}^{\infty}$ be a Markov Chain with, possibly empty, transient class $\mathcal{Y}$, and recurrent communicating classes $\mathcal{Z}_i$, $i = 1, 2, \ldots$. Then $\{Y(t)\}_{t=0}^{\infty}$ is *stable* if

$$\mathrm{Prob}\left[\min\{\tau \geq 0 \colon Y(\tau) \notin \mathcal{Y}\} < \infty \mid Y(0) = y\right] = 1, \quad \forall y \in \mathcal{Y},$$

and all states $z \in \cup_{i=1}^{\infty} \mathcal{Z}_i$ are positive recurrent.

We will say that the network is stable if the state process $\{\mathbf{X}(t)\}_{t=0}^{\infty}$ is stable, as defined in Definition 5.1.

5.3 A Broad Class of Policies Under Channel State Uncertainty

In this section, we introduce a general class of policies, $\mathcal{E}$. Our objective will be to compare the performance of the members in $\mathcal{E}$ to $\boldsymbol{\pi}_0^{\mathbf{w}}$ with respect to maximizing the stable throughput region of the network. This comparison will be performed in Section 5.4.

In order to specify the class $\mathcal{E}$ we define $n_{\hat{\mathbf{G}}\mathbf{E}\mathbf{Q}}(t;k,\mathbf{c},\mathbf{Q})$ to be the number of time slots in the interval $[0,t]$ that the estimated channel state is in state $\mathbf{G}^{(k)}$, the activation vector $\mathbf{E}(t)$ takes value $\mathbf{c}\in\Gamma_k$, and the matrix $\mathbf{Q}^{\mathbf{E}(t)}(t)$ is equal to $\mathbf{Q}\in\mathcal{Q}$. Here $\mathcal{Q}$ is the set of all $L\times L$ diagonal matrices whose diagonal is in the set $\{0,1\}^L$. Also, we define $n_{\hat{\mathbf{G}}\mathbf{E}}(t;k,\mathbf{c})$ to be the number of time slots in the interval $[0,t]$ that the estimated channel state is $\mathbf{G}^{(k)}$ and the activation vector $\mathbf{E}(t)$ takes value $\mathbf{c}\in\Gamma_k$. We define the set $\mathcal{E}$ as follows. We say that a policy $\{\mathbf{E}(t)\}_{t=1}^{\infty}$ belongs to $\mathcal{E}$ if for every $k,\ k'\in\mathcal{I}$ and time slot t the following is true

$$
\begin{aligned}
\mathrm{Prob}[\mathbf{G}(t) &= \mathbf{G}^{(k')}|\hat{\mathbf{G}}(t)=\mathbf{G}^{(k)},\mathbf{E}(t)=\mathbf{c}]\\
&= \mathrm{Prob}[\mathbf{G}(t)-\mathbf{G}^{(k')}|\hat{\mathbf{G}}(t)=\mathbf{G}^{(k)}],
\end{aligned}
\tag{5.19}
$$

and for every $k\in\mathcal{I}$, activation vector $\mathbf{c}\in\Gamma_k$, and matrix $\mathbf{Q}\in\mathcal{Q}$ the following is true

$$
\begin{aligned}
&\frac{n_{\hat{\mathbf{G}}\mathbf{E}\mathbf{Q}}(t;k,\mathbf{c},\mathbf{Q})}{n_{\hat{\mathbf{G}}\mathbf{E}}(t;k,\mathbf{c})}\\
&\to \frac{\mathrm{Prob}[\mathbf{Q}^{\mathbf{c}}(t)=\mathbf{Q},\hat{\mathbf{G}}(t)=\mathbf{G}^{(k)},\mathbf{E}(t)=\mathbf{c}]}{\mathrm{Prob}[\hat{\mathbf{G}}(t)=\mathbf{G}^{(k)},\mathbf{E}(t)=\mathbf{c}]}, \quad \text{a.s. as } t\to\infty
\end{aligned}
\tag{5.20}
$$

when $n_{\hat{\mathbf{G}}\mathbf{E}}(t;k,\mathbf{c})\neq 0$ as $t\to\infty$. Note that if $n_{\hat{\mathbf{G}}\mathbf{E}}(t;k,\mathbf{c})=0$ as $t\to\infty$, then the corresponding activation vector $\mathbf{c}$ is not used by the policy. In such a case, this activation vector can be eliminated from its constraint set. Recall that the constraint set is the set of all valid activation vectors with respect to the current channel state estimate.

Condition (5.19) is natural. It requires that at any time slot t, $\mathbf{E}(t)$ and the true channel state $\mathbf{G}(t)$ are conditionally independent given the estimate $\hat{\mathbf{G}}(t)$. In other words, all policies $\{\mathbf{E}(t)\}_{t=1}^{\infty}$ we may consider have no more information on the true channel state $\mathbf{G}(t)$ than the stationary policy $\boldsymbol{\pi}_0^{\mathbf{w}}$. Naturally, a policy that has additional information regarding the true channel state at time slot t can potentially exploit this knowledge and for example avoid collisions by not scheduling the corresponding nodes. Also, (5.20) is intuitive and is, in spirit, similar to a regular ergodicity condition. From (5.19) and (5.20) we may easily deduce that

$$\frac{n_{\hat{\mathbf{G}}\mathbf{EQ}}(t;k,\mathbf{c},\mathbf{Q})}{n_{\hat{\mathbf{G}}\mathbf{E}}(t;k,\mathbf{c})} \to \text{Prob}[\mathbf{Q}^{\mathbf{c}}(t) = \mathbf{Q} | \hat{\mathbf{G}}(t) = \mathbf{G}^{(k)}], \tag{5.21}$$

where from Assumption 2(a), $\text{Prob}[\mathbf{Q}^{\mathbf{c}}(t) = \mathbf{Q} | \hat{\mathbf{G}}(t) = \mathbf{G}^{(k)}]$ is independent of time t. Note also that the set $\mathcal{E}$ includes all the stationary policies since for stationary policies both (5.19) and (5.20) are being satisfied. It may further include some *non-stationary*, as well as *anticipative* policies as long as they comply with the conditions for being in set $\mathcal{E}$. Finally, we remind the reader that anticipative network control policies are all those policies that have knowledge on the future values of the quantities that affect the evolution of the state process driven by (5.6).

5.3.1 The Notion of Intermittent Boundedness

When the policy $\{\mathbf{E}(t)\}_{t=1}^{\infty}$ belongs to the class $\mathcal{E}$, the resulting queue size process $\{\mathbf{X}(t)\}_{t=0}^{\infty}$ generated by (5.6) is not necessarily a Markov Chain. Therefore, the stability definition according to Definition 5.1 is not applicable anymore. Instead, we will make use of a weaker notion of stability, that of intermittent boundedness.

Definition 5.2. The random process $\{Y(t)\}_{t=0}^{\infty}$ is almost surely intermittently bounded, if there exists a subset W of the sample space, with $\text{Prob}[W] = 1$, such that for every $\omega \in W$ there exists a sequence $\{t_i\}_{i=1}^{\infty}$ and a finite $Y_{\max}$ for which $|Y(\omega,t_i)| < Y_{\max}$, $\forall\, i = 1,2,\ldots$, where $Y(\omega,t)$ denotes the sample path of the process $\{Y(t)\}_{t=0}^{\infty}$ corresponding to outcome ω. Furthermore, $\{Y(t)\}_{t=0}^{\infty}$ is said to be intermittently bounded with positive probability, if there exists a subset W of the sample space, with $\text{Prob}[W] > 0$, such that for every $\omega \in W$ there exists a sequence $\{t_i\}_{i=1}^{\infty}$ and a finite $Y_{\max}$ for which $|Y(\omega,t_i)| < Y_{\max}$, $\forall\, i = 1,2,\ldots$.

5.4 Stable Throughput Optimality of the Introduced Class of Policies

In this section, we will present the optimality of the policy introduced in Section 5.2 with respect to maximizing the stable throughput region of the network under uncertainty in the channel state. We proceed to define some sets of rates that are important in the rest of this section.

In a stable network, traffic at any given node $n \in \mathcal{M}$ cannot accumulate without bound. Hence, stability can be viewed through the concept of *flow conservation*, namely that for every commodity the sum of departing flows at every node, except for the destination nodes for this commodity, must be equal to the sum of arriving flows for this commodity. Therefore, we define the set of *feasible* arrival rates $\mathbf{\Lambda}$ as

$$\mathbf{\Lambda} = \Bigg\{ \boldsymbol{\lambda} \in \mathbb{R}_{+}^{MH} : \exists \boldsymbol{\mu}_k^h \in \mathbb{R}_{+}^{L}, \quad \text{such that } \boldsymbol{\lambda}^h = -\mathbf{R}^h \sum_{k=1}^{K} p_{\hat{\mathbf{G}}}(k) \boldsymbol{\mu}_k^h \\ \text{and } \sum_{h=1}^{H} \boldsymbol{\mu}_k^h \in \text{co}(\tilde{\mathcal{Q}}_k) \Bigg\}, \tag{5.22}$$

where $\tilde{\mathcal{Q}}_k = \{\tilde{\mathbf{Q}}_k^{\mathbf{c}}\, \mathbf{c},\ \mathbf{c} \in \Gamma_k\}$, $\mathbf{f}_k^h$ are flow vectors of the hth commodity under estimated channel state $\mathbf{G}^{(k)}$ and $\text{co}(\cdot)$ denotes the convex hull of a set. Furthermore, let the stable, throughput region $\mathbf{C}_{\boldsymbol{\pi}_0^{\mathbf{w}}}$ under $\boldsymbol{\pi}_0^{\mathbf{w}}$ be defined as

$$\begin{aligned} \mathbf{C}_{\boldsymbol{\pi}_0^{\mathbf{w}}} = \Big\{ & \text{The set of arrival rates } \boldsymbol{\lambda} \text{ such that for all processes,} \\ & \big\{\mathbf{G}(t), \hat{\mathbf{G}}(t), \mathbf{A}(t)\big\}_{t=1}^{\infty} \text{ satisfying Assumptions 2 and 3,} \\ & \text{where } \boldsymbol{\lambda} = \mathbb{E}[\mathbf{A}(t)], \text{ the network is stable under } \boldsymbol{\pi}_0^{\mathbf{w}}.\Big\} \end{aligned}$$

We also denote by $\widetilde{\mathbf{C}}^1_{\boldsymbol{\pi}_0^{\mathbf{w}}}$ the following set of rates

$$\widetilde{\mathbf{C}}^1_{\boldsymbol{\pi}_0^{\mathbf{w}}} = \{\text{The set of rates } \boldsymbol{\lambda} \text{ such that for all processes } \{\mathbf{G}(t), \hat{\mathbf{G}}(t), \mathbf{A}(t)\}_{t=1}^{\infty}, \text{ satisfying Assumptions 2 and 3, where } \boldsymbol{\lambda} = \mathbb{E}[\mathbf{A}(t)], \text{ the process of the queue sizes is almost surely intermittently bounded under } \boldsymbol{\pi}_0^{\mathbf{w}}.\}$$

Finally, to compare with $\mathbf{C}_{\boldsymbol{\pi}_0^{\mathbf{w}}}$ and $\widetilde{\mathbf{C}}^1_{\boldsymbol{\pi}_0^{\mathbf{w}}}$, we introduce the set of arrival rates $\widetilde{\mathbf{C}}^p_{\mathcal{E}}$ as

$$\widetilde{\mathbf{C}}^p_{\mathcal{E}} = \{ \text{ The set of rates } \boldsymbol{\lambda} \text{ such that for some processes } \{\mathbf{G}(t), \hat{\mathbf{G}}(t), \mathbf{A}(t)\}_{t=1}^{\infty}, \text{ satisfying Assumption 2 where } \boldsymbol{\lambda} = \mathbb{E}[\mathbf{A}(t)], \text{ the process of the queue sizes is intermittently bounded with positive probability under some policy } \{\mathbf{E}(t)\}_{t=1}^{\infty} \in \mathcal{E}.\}$$

Note that although the requirement for an arrival rate being in $\mathbf{C}_{\boldsymbol{\pi}_0^{\mathbf{w}}}$ is that the process of the queue sizes is stable under $\boldsymbol{\pi}_0^{\mathbf{w}}$, the set of arrival rates $\widetilde{\mathbf{C}}^p_{\mathcal{E}}$ only requires that the queue size process satisfies the weak notion of intermittent boundedness with positive probability.

Let $\mathrm{ri}(\cdot)$ denote the relative interior of a set. The following theorem states the main result of optimality.

Theorem 5.2. The set $\boldsymbol{\Lambda}$ is a convex polytope. Furthermore, for *all* weight assignments $\mathbf{w} = (w_h, h = 1, 2, \ldots, H)$, with $w_h > 0$ for every commodity $h \in \mathcal{H}$, the following relationships hold

$$\mathrm{ri}(\boldsymbol{\Lambda}) \subseteq \mathbf{C}_{\boldsymbol{\pi}_0^{\mathbf{w}}} \subseteq \widetilde{\mathbf{C}}^1_{\boldsymbol{\pi}_0^{\mathbf{w}}} \subseteq \widetilde{\mathbf{C}}^p_{\mathcal{E}} \subseteq \boldsymbol{\Lambda}. \tag{5.23}$$

The proof of Theorem 5.2 appears in [52]. We proceed to give some more insight into the meaning of this theorem. From (5.23) it follows that for all weight assignments $\mathbf{w}$, the rate regions $\mathbf{C}_{\boldsymbol{\pi}_0^{\mathbf{w}}}$, $\widetilde{\mathbf{C}}^1_{\boldsymbol{\pi}_0^{\mathbf{w}}}$, and $\widetilde{\mathbf{C}}^p_{\mathcal{E}}$ are all squeezed between the convex polytope $\boldsymbol{\Lambda}$ and its relative interior. Hence, the sets of rates $\mathbf{C}_{\boldsymbol{\pi}_0^{\mathbf{w}}}$, $\widetilde{\mathbf{C}}^1_{\boldsymbol{\pi}_0^{\mathbf{w}}}$, and $\widetilde{\mathbf{C}}^p_{\mathcal{E}}$ can differ by at most points on the relative boundary of $\mathbf{\Lambda}$ and therefore they are almost identical sets. In fact, this implies that for any rate, except perhaps for a few rates in the relative boundary of $\mathbf{\Lambda}$, that cannot be stabilized by $\pi_0^{\mathbf{w}}$, there exists no policy in the large class $\mathcal{E}$ that can even make the process of the queue sizes intermittently bounded with some positive probability.

As an example, by utilizing (5.22), in Figure 5.2 we depict the stability region for the example network presented in Figure 5.1. Here, it is assumed that the uncertainty in the channel is such that the matrices $\tilde{\mathbf{Q}}_1^{[0,0]^T}$, $\tilde{\mathbf{Q}}_1^{[0,1]^T}$, $\tilde{\mathbf{Q}}_1^{[1,0]^T}$ are all equal to a diagonal matrix with diagonal entries given by 0.5, while the values of $\tilde{\mathbf{Q}}_2^{[0,0]^T}$ are immaterial due to the fact that there are no links available under channel state $\mathbf{G}^{(2)}$. Furthermore, we assumed that the stationary probabilities of the estimated channel states are both equal to 0.5, i.e., $p_{\hat{\mathbf{G}}}(1) = p_{\hat{\mathbf{G}}}(2) = 0.5$. As discussed above, the set of stable achievable rates belongs to the relative interior of $\mathbf{\Lambda}$ and thus differs from $\mathbf{\Lambda}$ in terms of the "relative boundary", as shown in Figure 5.2. Moreover, in Figure 5.2 we also provide the stability region of the network under perfect channel

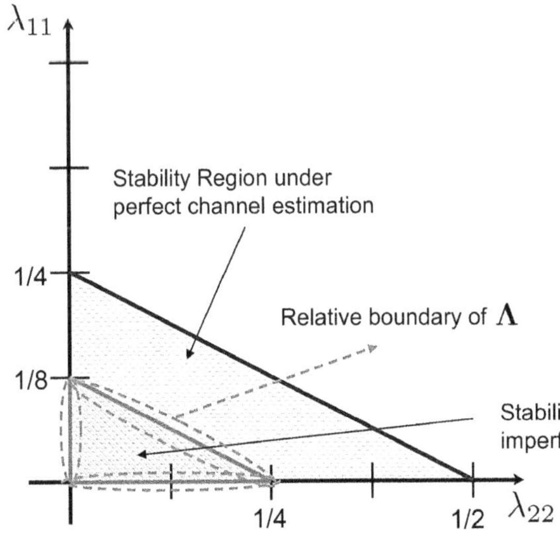

Fig. 5.2 Stability region of the network presented in Figure 5.1 under perfect and imperfect channel information.

information, obtained by replacing $\tilde{\mathbf{Q}}_1^{[0,0]^T}$, $\tilde{\mathbf{Q}}_1^{[0,1]^T}$, and $\tilde{\mathbf{Q}}_1^{[1,0]^T}$ with the identity matrix in (5.22). It is evident that the uncertainty in the channel state and the errors it introduces have a significant impact on the stability region.

6

Conclusions

As can be seen from the preceding sections, the problem of scheduling is very rich and complex and, yet, constitutes the foundation of how to operate a wireless network. It combines the issue of allocation of resources (bandwidth and power) with the aspects of physical layer signal design (modulation, coding, etc.) and, ultimately, with the routing and flow-control problems that synthesize the total picture of design and operation of multi-user systems in the wireless medium. It is also affected by the channel quality and by the performance criteria that dominate in each case. There cannot be a total and comprehensive solution unless unlikely breakthroughs arise from totally novel ideas or approaches. The best of current attempts to resolve the many thorny aspects of scheduling still result in combinatorially complex or analytically intractable formulations. Yet, through the systematic study of sub-problems that arise from these attempts, it is possible to "chip away" at the formidable challenge that the overall problem represents. We believe that the very selective review in this volume that, admittedly, leaves many aspects of the problem unaddressed, acquaints the reader with (i) the global version of scheduling and (ii) a detailed and in-depth exposition of some of our own recent work that illuminates some of the aspects of the scheduling problem. The latter represents a somewhat biased sampling of approaches to scheduling. However, it has the virtue of combining the physical layer with the resource allocation aspects of the media access control layer and thus amplifies the "layerless" view of wireless networking, that considers the parameters of different layers simultaneously and that is emerging as a dominant and useful philosophical approach to network analysis and design. Furthermore, it highlights the conceptual dichotomy of either pursuing heuristic sub-optimal solutions or optimizing over reduced strategy spaces. In addition, it offers insights and sometimes fairly general conclusions, like the optimality of solution structures under uncertainty. Undoubtedly, there is much more that needs to be done and we hope this volume inspires additional research in this area.

References

[1] N. Abramson, "The aloha system — another alternative for computer communications," in *Proceedings of AFIPS Fall Joint Computer Conference*, November 1970.

[2] R. Agrawal, A. Bedekar, R. J. La, and V. Subramanian, "Class and channel condition based weighted proportional fair scheduler," in *Proceedings of International Telegraffic Congress (ITC)*, 2001.

[3] R. Agrawal and V. Subramanian, "Optimality of certain channel aware scheduling policies," in *Proceedings of 40th Annual Allerton Conference on Communication, Control, and Computing*, September 2002.

[4] I. F. Akyildiz, J. McNair, L. C. Martorell, R. Puigjaner, and Y. Yesha, "Medium access control protocols for multimedia traffic in wireless networks," *IEEE Network*, vol. 13, no. 4, July/August 1999.

[5] P. Baran, "On distributed communication networks," *IEEE Transactions on Communications*, vol. 12, no. 1, March 1964.

[6] F. Berggren and R. Jantti, "Multiuser scheduling over rayleigh fading channels," in *Proceedings of IEEE Globecom*, December 2003.

[7] D. Bertsekas, *Dynamic Programming and Optimal Control, Vol. 1.* Athena Scientific, 3rd Edition, 2005.

[8] D. P. Bertsekas, *Dynamic Programming and Optimal Control, Vol. 2.* Athena Scientific, 3rd Edition, 2007.

[9] D. Bertsimas and J. N. Tsitsiklis, *Introduction to Linear Optimization.* Athena Scientific, 1997.

[10] S. A. Borbash, "Design considerations in wireless sensor networks," PhD thesis, Univ. of Maryland, 2004.

[11] S. A. Borbash and A. Ephremides, "The feasibility of matchings in a wireless network," *IEEE Transactions on Information Theory*, vol. 52, no. 6, June 2006.

[12] S. A. Borbash and A. Ephremides, "Wireless link scheduling with power control and SINR constraints," *IEEE Transactions on Information Theory*, vol. 52, no. 11, November 2006.

[13] S. Boyd and L. Vandenberghe, *Convex Optimization.* Cambridge University Press, 2004.

[14] P. Brémaud, *Markov Chains: Gibbs Fields, Monte Carlo Simulation and Queues.* Springer-Verlag, 1999.

[15] N. B. Chang and M. Liu, "Optimal channel probing and transmission scheduling for opportunistic spectrum access," *IEEE/ACM Transactions on Networking*, vol. 17, no. 6, December 2009.

[16] P. Chaporkar and S. Sarkar, "Minimizing delay in loss-tolerant MAC layer multicast," *IEEE Transactions on Information Theory*, vol. 52, no. 10, October 2006.

[17] T. H. Cormen, C. E. Leiserson, R. L. Rivest, and C. Stein, *Introduction to Algorithms.* The MIT Press, 2001.

[18] A. Ephremides and T. V. Truong, "Scheduling broadcasts in multihop radio networks," *IEEE Transactions on Communications*, vol. 38, no. 4, April 1990.

[19] F. Etemadi and H. Jafarkhani, "A unified framework for layered transmission over fading and packet erasure channels," *IEEE Transactions on Communications*, vol. 56, no. 4, 2008.

[20] D. D. Falconer, F. Adachi, and B. Gudmudson, "Time division multiple access methods for wireless personal communications," *IEEE Communications Magazine*, January 1995.

[21] G. Fayolle, P. Flajolet, M. Hofri, and P. Jacquet, "Analysis of a stack algorithm for random multiple-access communication," *IEEE Transactions on Information Theory*, vol. 31, no. 2, March 1985.

[22] R. Gallager, "A perspective on multiaccess channels," *IEEE Transactions on Information Theory*, vol. 31, no. 2, March 1985.

[23] L. Georgiadis, M. J. Neely, and L. Tassiulas, "Resource allocation and cross-layer control in wireless networks," *Foundations and Trends in Networking*, vol. 1, no. 1, 2006.

[24] L. Georgiadis, W. Szpankowski, and L. Tassiulas, "Stability analysis of quota allocation access protocols in ring networks with spatial reuse," *IEEE Transactions on Information Theory*, vol. 43, no. 3, May 1997.

[25] A. Goldsmith, *Wireless Communications.* Cambridge University Press, 2005.

[26] B. Hajek and G. Sasaki, "Link scheduling in polynomial time," *IEEE Transactions on Information Theory*, vol. 34, no. 5, September 1988.

[27] J. Hui and P. Humblet, "The capacity region of the totally asynchronous multiple-access channel," *IEEE Transactions on Information Theory*, vol. 31, no. 2, 1985.

[28] M. Johansson and L. Xiao, "Cross-layer optimization of wireless networks using nonlinear column generation," *IEEE Transactions on Wireless Communications*, vol. 5, no. 2, February 2006.

[29] K. Kar, X. Luo, and S. Sarkar, "Throughput-optimal scheduling in multichannel access point networks under infrequent channel measurements," *IEEE Transactions on Wireless Communications*, vol. 7, no. 7, July 2008.
[30] F. P. Kelly, "Charging and rate control for elastic traffic," *European Transactions on Telecommunications*, vol. 8, no. 1, 1997.
[31] F. P. Kelly, A. K. Maulloo, and D. K. H. Tan, "Rate control for commmunication networks: Shadow prices, proportional fairness and stability," *Journal of the Operational Research Society*, vol. 49, p. 1998.
[32] L. Kleinrock, "On queueing problems in random-access communications," *IEEE Transactions on Information Theory*, vol. 31, no. 2, March 1985.
[33] S. Kompella, J. E. Wieselthier, and A. Ephremides, "A cross-layer approach to optimal wireless link scheduling with SINR constraints," in *Proceedings of IEEE Military Communications Conference (MILCOM)*, October 2007.
[34] S. Kompella, J. E. Wieselthier, and A. Ephremides, "Revisiting the optimal scheduling problem," in *Proceedings of Conference on Information Sciences and Systems (CISS)*, March 2008.
[35] S. Kumar, V. S. Raghavan, and J. Deng, "Medium access control protocols for ad hoc wireless networks: A survey," *Elsevier Ad-Hoc Networks Journal*, vol. 4, no. 3, March 2006.
[36] M. S. Kuran, G. Gür, T. Tugcu, and F. Alagöz, "Applications of the cross-layer paradigm for improving the performance of WiMAX," *IEEE Wireless Communications*, vol. 17, no. 3, June 2010.
[37] H. J. Kushner and P. A. Whiting, "Asymptotic properties of proportional-fair sharing algorithms," in *Proceedings of 40th Annual Allerton Conference on Communication, Control, and Computing*, September 2002.
[38] H. J. Kushner and P. A. Whiting, "Convergence of proportional-fair sharing algorithms under general conditions," *IEEE Transactions on Wireless Communications*, vol. 3, no. 4, 2004.
[39] A. Lapidoth and P. Narayan, "Reliable communication under channel uncertainty," *IEEE Transactions on Information Theory*, vol. 44, no. 6, October 1998.
[40] X. Lin, N. B. Shroff, and R. Srikant, "A tutorial on cross-layer optimization in wireless networks," *IEEE Journal on Selected Areas in Communications*, vol. 24, no. 8, August 2006.
[41] J. Massey and P. Mathys, "The collision channel without feedback," *IEEE Transactions on Information Theory*, vol. 31, no. 2, March 1985.
[42] S. Merlin, N. Vaidya, and M. Zorzi, "Resource allocation in multi-radio multi-channel multi-hop wireless networks," in *Proceedings of IEEE INFOCOM*, April 2008.
[43] J. Mo and J. Walrand, "Fair end-to-end window-based congestion control," *IEEE/ACM Transactions on Networking*, vol. 8, no. 5, 2000.
[44] M. J. Neely, "Max weight learning algorithms with application to scheduling in unknown environments," in *ITA 2009*, http://arxiv.org/abs/0902.0630v1.
[45] M. J. Neely, E. Modiano, and C. E. Rohrs, "Dynamic power allocation and routing for time-varying wireless networks," *IEEE Journal on Selected Areas in Communications*, vol. 23, no. 1, January 2005.

[46] A. Pantelidou and A. Ephremides, "A cross-layer view of wireless multicast optimization," in *Proceedings of 46th Annual Allerton Conference on Communication, Control, and Computing*, September 2008.
[47] A. Pantelidou and A. Ephremides, "Minimum schedule lengths with rate control in single-hop wireless networks," in *Proceedings of IEEE Military Communications Conference (MILCOM)*, November 2008.
[48] A. Pantelidou and A. Ephremides, "Optimal rate control policies for proportional fairness in wireless networks," in *Proceedings of Conference on Information Sciences and Systems (CISS)*, March 2008.
[49] A. Pantelidou and A. Ephremides, "What is optimal scheduling in wireless networks?," in *Proceedings of 4th International Wireless Internet Conference (WICON)*, November 2008.
[50] A. Pantelidou and A. Ephremides, "A cross-layer view of optimal scheduling," *IEEE Transactions on Information Theory*, vol. 56, no. 11, November 2010.
[51] A. Pantelidou, A. Ephremides, and A. Tits, "Maximum throughput scheduling in time-varying-topology wireless ad-hoc networks," in *Proceedings of Conference on Information Sciences and Systems (CISS)*, March 2005.
[52] A. Pantelidou, A. Ephremides, and A. L. Tits, "A cross-layer approach for stable throughput maximization under channel state uncertainty," *ACM/Kluwer Journal of Wireless Networks*, vol. 15, no. 5, July 2009.
[53] B. Radunović and J. L. Boudec, "Rate performance objectives of multi-hop wireless networks," in *Proceedings of IEEE INFOCOM*, March 2004.
[54] C. Raman, R. D. Yates, and N. B. Mandayam, "Scheduling variable rate links via a spectrum server," in *Proceedings of IEEE Dynamic Spectrum Access Networks (DySPAN)*, November 2005.
[55] L. G. Roberts, "Aloha packet system with and without slots and capture," *Newsletter ACM SIGCOMM Computer Communication Review*, vol. 5, no. 2, 1975.
[56] A. Sadek, K. J. R. Liu, and A. Ephremides, "Cognitive multiple access via cooperation: Protocol design and performance analysis," *IEEE Transactions on Information Theory*, vol. 53, no. 10, October 2007.
[57] S. Shakkottai, T. S. Rappaport, and P. Karlsson, "Cross-layer design for wireless networks," *IEEE Communications Magazine*, vol. 41, no. 10, October 2003.
[58] S. Shakkottai and A. Stolyar, "Scheduling for multiple flows sharing a time-varying channel: The exponential rule," *Analytic Methods in Applied Probability. In Memory of Fridrih Karpelevich. Yu. M. Suhov, Editor. American Mathematical Society Translations, Series*, vol. 207, no. 2, 2000.
[59] G. Sharma, A. Ganesh, and P. Key, "Performance analysis of contention based medium access control protocols," *IEEE Transactions on Information Theory*, vol. 55, no. 4, April 2009.
[60] B. Shrader and A. Ephremides, "On the queueing delay of a multicast erasure channel," in *Proceedings of IEEE Information Theory Workshop (ITW)*, October 2006.
[61] J. So and N. Vaidya, "Multi-channel MAC for ad hoc networks: Handling multichannel hidden terminals using a single transceiver," in *Proceedings of ACM MOBIHOC*, May 2004.

[62] A. Somekh-Baruch, S. Shamai, and S. Verdú, "Cooperative multiple-access encoding with states available at one transmitter," *IEEE Transactions on Information Theory*, vol. 54, no. 10, October 2008.
[63] V. Srivastava and M. Motani, "Cross-layer design: A survey and the road ahead," *IEEE Communications Magazine*, vol. 43, no. 12, December 2005.
[64] A. L. Stolyar, "On the asymptotic optimality of the gradient scheduling algorithm for multiuser throughput allocation," *Operations Research*, vol. 53, no. 1, January–February 2005.
[65] L. Tassiulas, "Scheduling and performance limits of networks with constantly changing topology," *IEEE Transactions on Information Theory*, vol. 43, no. 3, May 1997.
[66] L. Tassiulas and A. Ephremides, "Stability properties of constrained queueing systems and scheduling policies for maximum throughput in multihop radio networks," *IEEE Transactions on Automatic Control*, vol. 37, no. 12, December 1992.
[67] L. Tassiulas and A. Ephremides, "Dynamic server allocation to parallel queues with randomly varying connectivity," *IEEE Transactions on Information Theory*, vol. 39, no. 2, March 1993.
[68] V. Tsibonis, L. Georgiadis, and L. Tassiulas, "Exploiting wireless channel state information for throughput maximization," *IEEE Transactions on Information Theory*, vol. 50, no. 11, November 2004.
[69] M. van Der Schaar and N. S. Shankar, "Cross-layer wireless multimedia transmission: challenges, principles, and new paradigms," *IEEE Wireless Communications*, vol. 12, no. 4, August 2005.
[70] D. Vasudevan, V. G. Subramanian, and D. J. Leith, "On ARQ for packet erasure channels with bernoulli arrivals," in *Proceedings of IEEE International Symposium on Information Theory (ISIT)*, June 2010.
[71] S. Verdú, *Multiuser Detection*. Cambridge University Press, 2003.
[72] S. Vishwanath, N. Jindal, and A. Goldsmith, "Duality, achievable rates, and sum-rate capacity of gaussian mimo broadcast channels," *IEEE Transactions on Information Theory*, vol. 49, no. 10, 2003.
[73] X. Wang, G. B. Giannakis, and A. G. Marques, "A unified approach to QoS-guaranteed scheduling for channel-adaptive wireless networks," *Proceedings of the IEEE*, vol. 95, no. 12, 2007.
[74] H. Won, H. Cai, D. Y. Eun, K. Guo, A. Netravali, I. Rhee, and K. Sabnani, "Multicast scheduling in cellular data networks," in *Proceedings of IEEE INFOCOM*, May 2007.
[75] P. Xia, S. Zhou, and G. B. Giannakis, "Adaptive MIMO OFDM based on partial channel state information," *IEEE Transactions on Signal Processing*, vol. 52, no. 1, January 2004.
[76] L. Ying and S. Shakkottai, "Scheduling in mobile wireless networks with topology and channel-state uncertainty," in *Proceedings of IEEE INFOCOM*, 2009.

Lightning Source UK Ltd.
Milton Keynes UK
UKOW05f1103231216
290740UK00001B/21/P

9 781601 984203